CONTE   QUE   ONS FOR
ADVANCED
CHEMISTRY

B. Earl
B.Sc., P.G.C.E., C.Chem., M.R.S.C. Head of Chemistry,
St Aidan's Church of England High School, Harrogate

L.D.R. Wilford
B.Tech., PhD., P.G.C.E., C.Chem., F.R.S.C. Head of Science Faculty,
St Aidan's Church of England High School, Harrogate

Nelson Blackie

CONTEXT QUESTIONS FOR ADVANCED CHEMISTRY

First published 1992
Reprinted 1993

ISBN 017448224-8
NPN 10 9 8 7 6 5 4 3

A catalogue of this book is available from The British Library

Front cover courtesy of Science Photo Library

Published by
Thomas Nelson and Sons Ltd
Nelson House Mayfield Road
Walton-on-Thames Surrey
KT12 5PL UK

51 York Place
Edinburgh
EH1 3JD UK

Nelson Blackie
Wester Cleddens Road
Bishopbriggs
Glasgow G64 2NZ UK

Thomas Nelson (Hong Kong) Ltd
Toppan Building 10/F/22A Westlands Road
Quarry Bay Hong Kong

Thomas Nelson Australia
102 Dodds Street
South Melbourne
Victoria 3205 Australia

Nelson Canada
1120 Birchmount Road
Scarborough Ontario
M1K 5G4 Canada

Authors' Acknowledgements
The authors wish to thank Helen Fenton for her comments; Dennis Richards,
Headmaster, St Aiden's Church of England High School, Harrogate,
for his help and support; and Irene and Barbara for their patience and
encouragement throughout.

Printed and bound in Great Britain by Bell and Bain Ltd, Glasgow

CONTENTS

The figure in brackets is the page number on which the corresponding answers to a question will be found.

	Introduction	(v)	
1	CFC12	1	[47]
2	An Organic Problem	2	[47]
3	Industrial Equilibrium	3	[47]
4	Fission/Fusion - 1	4	[48]
5	Equations of State	5	[48]
6	A Friedel-Crafts Catalyst	6	[49]
7	An Essential Metal	7	[49]
8	A Genetic Element	8	[50]
9	Analysis of a Mineral Water	9	[50]
10	Fission/Fusion - 2	11	[51]
11	Biochemistry	12	[52]
12	Graphitic Compounds	13	[52]
13	Effusion	14	[53]
14	Relative Molecular Mass	15	[53]
15	Anti-Cancer Drug	16	[53]
16	Iron	17	[54]
17	An Analgesic	19	[54]
18	Titanium	20	[55]
19	Reactions of Anaesthetics	21	[55]
20	Tackling Sleeping Sickness	22	[56]
21	Enthalpy of Combustion	23	[56]
22	Reaction Kinetics	24	[56]
23	Newly Discovered Elements	25	[57]
24	Sulphuric Acid	26	[57]
25	Ammonia	27	[57]
26	Chloride Analysis	29	[58]
27	Household Bleaches	30	[58]
28	Aspirin	31	[58]
29	A Useful Salt	33	[59]
30	A Versatile Transition Metal	34	[60]
31	Natural Gas	35	[60]
32	Internal Combustion Engine	36	[60]
33	Health Drink	37	[61]
34	Chemical Rockets	38	[61]
35	North Sea Analysis	39	[61]
36	Water	40	[62]
37	Chlorine in the Environment	41	[62]
38	Forensic Science	42	[63]
39	Lead Pollution	43	[63]
40	Corrosion	44	[63]
	Answers	47	
	Graphs	64	
	Periodic Table	66	

INTRODUCTION

This book contains forty photocopiable context-set questions for Advanced and Advanced Supplementary courses in Chemistry. The questions have been written in a language accessible to the full range of students at this level and are designed to be used as part of the day to day teaching and as such could then form part of the students' own notes. Each context question is accompanied by a complete set of answers.

In the matter of units and nomenclature, we have been guided by the advice of the Examination Boards and by the relevant publications of the Association for Science Education.

B. Earl and D. Wilford

1   CFC 12

Chlorofluorocarbons (CFCs) are powerful greenhouse gases. There are also fears over the impact of CFCs on the ozone layer. Ozone depletion has become such a serious problem that an international agreement known as the 'Montreal Protocol on Substances Which Deplete the Ozone Layer' was agreed in 1987. The proposed controls were tightened in 1990 by the second meeting of the parties to the 'Montreal Protocol'.

Read through the sequence below which outlines the reactions used in the formation of CFC 12; dichlorodifluoromethane (CCl_2F_2).

$$3CCl_{4(l)} + 2X_{(s)} \longrightarrow 3CCl_2F_{2(g)} + 2SbCl_{3(s)}$$

X is then regenerated using liquid hydrogen fluoride:

$$6 \text{ hydrogen fluoride}_{(l)} + 2SbCl_{3(s)} \longrightarrow 2X_{(s)} + 6Z_{(g)}$$

1 Identify the materials X and Z by name and formula.

2 Complete the chemical equations for the reactions shown above.

3 Explain the meaning of the terms 'greenhouse gases' and 'ozone depletion'

4 Give two former uses of CFC 12.

The historic and predicted concentrations of CFC 12 are shown on the graph opposite. ('Now' refers to the year of preparation of the figures, that is, 1989).

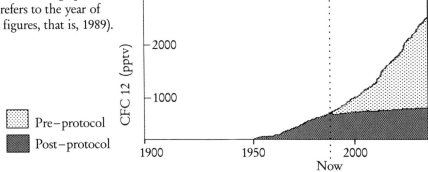

5 Discuss the way that the concentration of CFC 12 would vary between 1989 and 2050,

    a Pre–protocol,

    b Post–protocol.

CFC 12 is almost chemically inert, for example it will not react even with molten sodium. Because of its inertness there is no natural process for removing CFC 12 from the environment. CFC 12 drifts up into the stratosphere where a photochemical reaction takes place and chlorine free radicals are produced:

$$CFC\ 12 \xrightarrow{\text{uv}} {}^{\bullet}A + {}^{\bullet}CF_2$$

Free radicals then react with ozone and convert it into oxygen:

$$\text{ozone} + {}^{\bullet}A \longrightarrow D + O_2$$

$$\text{then} \qquad D + {}^{\bullet}O \longrightarrow {}^{\bullet}A + O_2$$

$$\text{also} \qquad D + O_3 \longrightarrow ClO_2 + O_2$$

6 Identify each of the species A and D by name and formula.

7 Complete and balance the chemical equations involving species A and D.

8 Study the sequence of reactions involving A and D. Why do you think the formation of one chlorine free radical is so dangerous ?

9 Use the electron pair repulsion theory to deduce the shape and bond angles of CFC 12.

2   An Organic Problem

The following data relate to an organic substance X.

Physical state at room temperature	Liquid	
Relative molecular mass	70	
Elemental analysis (% by mass)	C (68.57%), H (8.57%), O (22.86%)	
Chemical tests	Reagent	Results
A	2,4–dinitrophenylhydrazine	Orange precipitate
B	Ammoniacal silver nitrate	No reaction
C	Bromine (dissolved in 1,1,1–trichloroethane)	Decolourised

Use these data to answer the following questions.

1 What is the empirical formula of X ?
 (R.A.M., A_r C=12, H=1, O=16)

2 What is the molecular formula of X ?

3 What can be said about the structure of X by considering :
 a test A alone,
 b tests A and B taken together,
 c test C alone ?

4 On the basis of your answers to questions **2** and **3** suggest a possible structure for X.

5 Use the electron pair repulsion theory to deduce the approximate bond angles present within this molecule.

6 a Explain what is meant by the term 'polar bond'.
 b Identify a polar bond within the molecule of X.

7 a Write a balanced chemical equation for the reaction taking place in test C.
 b State the mechanism for this reaction.

3 Industrial Equilibrium

Gaseous ammonia is manufactured in large quantities by the Haber Process. This process was developed by Fritz Haber in 1911. The production of this important chemical is affected by both the temperature and pressure at which the process is run. The equation which represents the synthesis of ammonia is

$$N_{2(g)} + 3H_{2(g)} \rightleftharpoons 2NH_{3(g)}$$

A catalyst of iron is used with aluminium oxide as a promoter.

The graph shows how the percentage yield of ammonia changes with temperature and pressure.

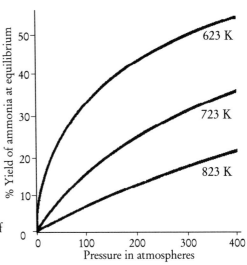

1 Using the graphical information state :
  a i the qualitative effect of increased pressure,
     ii the qualitative effect of increased temperature,
        on the percentage yield of ammonia.
  b Using your answer to **a ii** deduce the sign of and give an explanation of the enthalpy change which occurs during the production of ammonia.

2 a The conditions employed in industry for the production of ammonia are commonly of the order of 200 atmospheres pressure at a temperature of 723 K. What is the percentage yield of ammonia using these conditions?
  b The following data were obtained from a trial run of ICI's ammonia plant at Billingham using the conditions specified in part **2a** :
       200 000 dm³ of nitrogen were mixed with 700 000 dm³ of hydrogen in the presence of an iron catalyst. Using your answer to part **2a** calculate the volume and mass of ammonia produced under these conditions.
  c The percentage yield of ammonia is greater than 80% if the process is run at 400 atmospheres and 450 K. Why are these conditions not employed ?

3 The boiling points of the three gases present in the equilibrium mixture are

Gas	Boiling Point/K
Nitrogen	77
Hydrogen	20
Ammonia	240

  a Explain how the ammonia produced in the Haber process can be separated from the unchanged nitrogen and hydrogen.
  b Why is the ammonia removed from the equilibrium mixture?

4 a Assuming that nitrogen and hydrogen gases react in the mole ratios shown in the balanced equation, at 773 K and 200 atmospheres pressure, and that a 20% yield of ammonia is obtained in the equilibrium mixture, calculate the percentage of:
       i nitrogen,
      ii hydrogen in the equilibrium mixture.
  b Write an expression for K_p for the Haber Process.
  c Calculate the partial pressures of the three gases in the equilibrium mixture and hence a value for K_p.
  d Would you expect the value of K_p to increase or decrease at a lower pressure ?
  e What assumptions have you made about the three gases involved in calculating K_p ?

5 a What is the purpose of using the iron catalyst in this process ?
  b Draw a suitable energy level diagram showing the effect of this catalyst on the reaction mechanism of the Haber Process.

Uranium was discovered by E–M. Peligot in 1842 in Paris, France. Until the discovery of uranium fission by Hahn and Strassman in 1939, uranium had little commercial importance. Uranium is now a very important nuclear fuel.

The most important of the widely distributed sources of uranium is pitchblende, a mixed oxide. The extraction of uranium from this ore involves initial reaction with nitric acid followed by thermal decomposition of the nitrate product.

The remaining stages of the process can be summarised in the reaction scheme below. Read through this sequence which outlines the final series of reactions used in the extraction of uranium.

$$UO_{3(s)} \; + \; A_{(g)} \longrightarrow D_{(s)} \; + \; H_2O_{(g)}$$

$$D_{(s)} \; + \; 4\,HF_{(g)} \longrightarrow UF_{4(s)} \; + \; H_2O_{(g)}$$

$$UF_{4(s)} \; + \; Mg_{(s)} \longrightarrow U_{(s)} \; + \; E_{(s)}$$

1 Identify the materials A, D and E by name and formula. Complete and balance the chemical equations for the three reactions of the sequence.

2 a Calculate the number of moles of uranium obtained in the final stage of this sequence from 471 kg of uranium(IV) fluoride in reaction with an excess of magnesium.

   b The natural abundances of the two isotopes of uranium are ^{235}U 0.72% and ^{238}U 99.28%. Calculate the mass of ^{238}U produced in this process. (R.A.M., A_r F=19)

3 Suggest a metal which could be used instead of magnesium to reduce UF_4 to the metal. Justify your choice of metal.

4 When pitchblende is dissolved in nitric acid a yellow salt, $UO_2(NO_3)_2.6H_2O$ is obtained. Give the oxidation state of uranium: **a** in this salt,

              b in UO_3.

Uranium has properties of a typical metal. For example, it has a body–centred cubic structure as well as having an electrical conductivity which is similar to that of other dense metals such as lead.

5 Sketch a body–centred cubic unit cell such as that found in uranium.

6 Explain how uranium is capable of conducting electricity.

7 Uranium is also quite a chemically reactive metal. For example, it will react with:

  a boiling water giving uranium(IV) oxide and hydrogen,

  b hydrogen producing uranium(III) hydride.

Write a balanced chemical equation for each of these reactions.

5 Equations of State

British Gas is one of many industries interested in the gas laws which are often called the 'equations of state'.

The simplest example of this type of equation of state is the ideal gas equation:

$$pV = nRT$$

where: p = pressure, V = volume and T = temperature.

1 Define any terms not already described in the equation above.

The ideal gas equation is used as a starting point for an equation used to model the behaviour of real gases. Van der Waals made small modifications to this equation in 1873 to produce the well known van der Waals' real gas equation:

$$(P + a/V^2)(V - b) = nRT$$

These equations are central to the modelling which takes place in aspects of materials processing within this industry. Using modifications of these equations, predictions can be made which have a direct relevance to the functioning and economy of gas treatment plants, gas pipelines and gas reservoirs.

2 What is the significance of the terms 'a/V^2' and '$V- b$' as shown in the van der Waals' equation ?

3 Why does the real gas equation reduce to $pV=RT$ for one mole of any gas at low pressures and high temperatures ?

A British Gas pilot plant reservoir holds 1×10^5 dm^3 of methane.

4 a At a temperature of 25 °C, calculate the pressure produced when 1000 kg of methane is introduced into the reservoir by using :

  i the ideal gas equation,

  ii the van der Waals' real gas equation.

  (R.A.M., A_r C=12, H=1. Van der Waals' constants for methane a = 2.25 atm dm^6 mol^{-2} and

  b = 0.043 dm^3 mol^{-1} and R = 0.0821 dm^3 atm K^{-1} mol^{-1}).

  b Account for the different values obtained using the two equations.

5 a Draw a diagram to show the bonding present in methane.

  b Use the electron pair repulsion theory to explain the shape of the methane molecule.

6 A Friedel–Crafts Catalyst

Friedel–Crafts catalysts are named after C.Friedel and J.M.Crafts who discovered them. Aluminium halides are examples of Friedel–Crafts catalysts and may be prepared by direct synthesis. The following data relate to a particular experiment to prepare aluminium chloride directly. Also included in the table are details of some of the properties of this substance.

Preparation	Dry chlorine was passed over 0.54g of heated aluminium. 2.67g of aluminium chloride (a white powder) were collected under anhydrous conditions.
Relative molecular mass	267 (solid). 133.5 (in benzene solution).
Exposure of aluminium chloride to air	Misty fumes were formed which turned moist indicator paper red.
Reaction with water	An exothermic reaction occured and an acidic solution was produced.

1 Explain why anhydrous conditions are used in the preparation of aluminium chloride.

2 Using the relative molecular mass data, determine the formulae for both aluminium chloride as a solid and in benzene solution. (RAM., A_r Al=27, Cl=35.5)
   Comment on any difference noted.

3 Name and account for the formation of the substance responsible for the misty fumes produced when aluminium chloride is exposed to air.

4 a Give the formula of the principal aluminium–containing complex ion present in the colourless solution formed when aluminium chloride is added to an excess of water.
   b Account for the acidity of this solution.

5 Anhydrous aluminium chloride will act as a Friedel–Crafts catalyst in the reaction between ethanoyl chloride and benzene.
   a What feature of aluminium chloride allows it to act as a catalyst in this reaction ?
   b Write a balanced chemical equation to represent the reaction between ethanoyl chloride and benzene.
   c State the mechanism of this reaction.

7   An Essential Metal

Metallic element X was discovered in 1808 by Sir H. Davy, in London. It is used as a reducing agent in the extraction of metals such as vanadium as well as being used in the manufacture of some alloys. The following data relate to element Y and some of its reactions.

Metal/ non−metal	Metal
Colour	Shiny but tarnishes on exposure to air.
Melting point/K	1112.0
Boiling point/K	1757.0
Flame colour	Brick red.
Reaction with water	When cleaned thoroughly with abrasive paper, element X reacts quite vigorously. An alkaline solution A and a colourless gas D are produced.
Reaction with oxygen	When cleaned thoroughly with abrasive paper and heated strongly, element X burns very vigorously with a brick red flame. Clouds of white smoke are produced which settle as a white deposit E.
Reaction with chlorine	When cleaned thoroughly with abrasive paper and heated stongly in a stream of chlorine, element X burns strongly. Clouds of white smoke are produced which settle as a white deposit G.

Use these data to answer questions 1 to 5.

1 What is the physical state of X at room temperature ?

2 Name the element X and give its:
   a symbol,   **b** relative atomic mass,
   c atomic number,   **d** electronic configuration of the atom and its most common ion,
   e group number,   **f** period number.

3 **a** Write a balanced chemical equation for the reaction of element X with water and identify, by formula and name, both compound A and gas D.
   b Give the formulae of the ions present in the alkaline solution of A.

4 **a** Write a chemical equation for the formation of compound E and give its name and formula.
   b Compound E is used for drying gases such as ammonia. Write a chemical equation for the reaction which takes place between compound E and water.

5 **a** Write a chemical equation for the formation of compound G and give its name and formula.
   b Draw a diagram to show the bonding present in compound E.

6 In the solid state, metals have their atoms packed closely together in a three−dimensional array. Metal X has a face−centred cubic structure.
   a Draw a diagram to illustrate this type of packing.
   b This metal is a good conductor of electricity. Explain this property by reference to the type of bonding present in metals.

© Earl and Wilford 1992

8   A Genetic Element

The non–metallic element X was discovered in 1669 by H. Brand in Hamburg, Germany. X is essential to life as a constituent of cell membranes as well as some proteins and all nucleic acids and nucleotides.

The following data relate to the element X and some of its reactions.

Metal/non–metal	Non–metal
Colour	White
Melting point/K	317
Boiling point/K	533
Reaction with oxygen	Burns spontaneously on exposure to air. A white powder of empirical formula X_2O_5 is produced.
Reaction with chlorine	Burns spontaneously in chlorine. A mixture of liquid A and solid D is produced.
Reaction with hot aqueous sodium hydroxide	A gas E is produced which is only sparingly soluble in water.

Use these data to answer the following questions.

1 What is the physical state of X at room temperature ?

2 Name the element X and give its:
   a symbol,
   b relative atomic mass,
   c atomic number,
   d electronic configuration of the atom,
   e electronic configuration of one of its ions,
   f group number,
   g period number.

3 Write a balanced chemical equation for the reaction of element X with oxygen.

4 a Give the formula of compounds A and D, and write a balanced chemical equation for their formation.
   b Use the electron pair repulsion theory to describe the shape of compounds A and D.

5 Give the formula of compound E, and write a balanced chemical equation for its formation.

6 a Name three other elements found in substances essential for life, for example, in DNA.
   b For the elements you have named in part **a**, give by name and formula an example of a compound containing the elements in their highest oxidation state.

Mineral waters are usually taken from an underground source. To comply with 'The Natural Mineral Waters' regulations (1985), the water must be analysed. The following table of data shows a typical analysis of a spring water which is sold in many high street outlets.

Mineral Analysis	Typical Values / mg per dm^3
Calcium	22.0
Magnesium	5.0
Potassium	0.3
Sodium	6.8
Bicarbonate	77.0
Sulphate	12.0
Nitrate	2.0
Fluoride	0.1
Chloride	10.0
Silicate	5.7
Dry residue at 180°C	95.0
pH	7.2

1 Describe the chemical tests to show the presence of:

   a sulphate ion,

   b nitrate ion,

   c chloride ion,

      in this spring water.

      Detail any problems associated with carrying out these tests using this spring water.

2 Give the oxidation state of:

   a sulphur in the sulphate ion,

   b nitrogen in the nitrate ion,

   c fluorine in the fluoride ion,

   d silicon in the silcate ion.

3 Use the electron pair repulsion theory to explain the shape of:

   a the nitrate ion,

   b the sulphate ion.

   In your answer include the approximate bond angles.

4 Describe with full experimental details how you could determine the concentration of chloride ions (mg per dm^3) in a sample of this spring water.

5 a Give a possible explanation of the value obtained for the pH of this spring water.

   b Define pH and calculate the concentration of hydrogen ions in moles per dm^3 in the spring water.

6 The total mass of all the substances shown in the table is 140.90 mg. Explain why the dry residue, after heating to 180°C is only 95 mg from 1 dm^3 of spring water.

7 The metals present in the mineral analysis are given below along with some other data.

Metal	Group	Atomic Number	Electronic Configuration		Radius/nm		1st Ionisation Energy/kJ mol^{-1}
			Atom	Ion	Atomic	Ionic	
Sodium	1	11	$1s^2 2s^2$	$1s^2 2s^2$	0.186	0.098	+496
Potassium	1	19	$1s^2 2s^2$	$1s^2 2s^2$	0.227	0.133	+419
Magnesium	2	12	$1s^2 2s^2$	$1s^2 2s^2$	0.160	0.065	+738
Calcium	2	20	$1s^2 2s^2$	$1s^2 2s^2$	0.197	0.100	+590

    a Complete the electronic configurations for the atoms of these metals as well as those for their most stable ions.

    b Account for the trends in atomic and ionic radii as well as first ionisation energies:

      i within the groups indicated,

      ii going across the periods containing the metal elements indicated in the table.

    c What would you expect to happen to the value of the second ionisation energy in comparison with the first ionisation energy for the metal elements shown. Explain your answer.

10   Fission/Fusion – 2

In a gas cooled nuclear reactor it is the ^{235}U isolated from natural uranium that undergoes <u>fission</u>. Eventually the fuel elements need replacing because so much of the ^{235}U has been turned into other products:

$$^{235}_{92}U + {}^{1}_{0}n \longrightarrow {}^{92}_{a}X + {}^{b}_{56}E + 3\,{}^{1}_{0}n$$

1 Explain the meaning of the symbol $^{235}_{92}$U.

2 Explain what is meant by the term nuclear fission.

3 Complete the nuclear equation, identifying the elements X and E as well as calculating their atomic number and mass number.

^{238}U does not undergo fission. However, it can capture a neutron turning it into ^{239}U. ^{239}U then decays producing a <u>daughter element</u> G which then decays to ^{239}Pu.

$$^{238}_{92}U + {}^{1}_{0}n \longrightarrow {}^{239}_{92}U \longrightarrow G \longrightarrow {}^{239}_{94}Pu$$

4 Explain what is meant by the term 'daughter element'.

5 Write full nuclear equations to represent the processes taking place in the radioactive decay of ^{239}U.

6 Name the type of reactor which makes use of the plutonium produced by this process.

The energy released from these nuclear reactions is removed by a coolant such as carbon dioxide or water at high pressure.

7 How can electricity be generated using these 'hot' coolant substances ?

There is a further way of obtaining energy from nuclei, that is, by <u>fusion</u> of hydrogen nuclei. This process can be represented by the nuclear equation:

$$^{2}_{1}H + {}^{2}_{1}H \longrightarrow {}^{c}_{d}J + {}^{1}_{0}n$$

Explain what is meant by the term 'fusion'.

9 Name the isotope of hydrogen used in this process.

10 Identify the element J along with its mass and atomic numbers c and d respectively

11 Biochemistry

Glycine is an amino acid. It is a white solid with a melting point of 232°C. It can be made from ethanoic acid by the following reaction sequence.

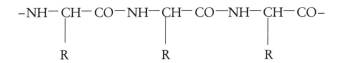

$$CH_3COOH \xrightarrow{\quad a \quad} ClCH_2COOH \xrightarrow{\quad b \quad} H_2NCH_2COOH$$

1 State the reagents and conditions needed to carry out each of the steps *a* and *b*.

2 Describe the mechanism of the reactions taking place in:

  i step *a,*

  ii step *b.*

Polymerisation is the most important type of reaction of amino acids. Enzymes catalyse the polymerisation of amino acids in living organisms. The compounds produced are known as proteins or polypeptides. A part of a general protein structure is shown below:

$$-NH-\underset{\underset{R}{|}}{CH}-CO-NH-\underset{\underset{R}{|}}{CH}-CO-NH-\underset{\underset{R}{|}}{CH}-CO-$$

3 Name and draw the structure of the major type of linkage which is found within a protein molecule.

4 What is the difference between a polypeptide and a protein ?

5 What is the role of protein material in the control of metabolism ?

The formula below represents part of a synthetic polymer which contains the same type of linkage as that found in proteins, i.e.

$$-NH(CH_2)_6NHCO(CH_2)_4CONH(CH_2)_6NHCO(CH_2)_4CO-$$

6 What is the name of the synthetic polymer ?

7 Draw the structural formulae of the monomers which could be used to produce this polymer.

8 Name a reagent which could be used to hydrolyse this polymer.

9 The polymer you have named in question **6** can be drawn into fibres. For a polymer to have fibre properties there needs to be relatively strong attractive forces between adjacent polymer chains.

  a Identify the group(s) of atoms responsible for this attraction.

  b Name the type of interaction involved.

12　Graphitic Compounds

Graphite is one of the underlined{allotropes of carbon}. The graphite structure consists of layers and is shown below:

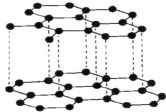

Within each layer, each carbon atom is bonded to three other carbon atoms by three strong covalent bonds. There are, however, only weak forces holding the layers together and so the layers will easily pass over each other. The unsaturation present in a graphite lattice is responsible for an unusual set of compounds called 'graphitic compounds'. In these compounds, different atoms can be fitted in between the layers of carbon atoms to produce compounds with a non–stoichiometric composition. When the graphite lattice contains atoms of alkali metals such as caesium, then a bronze–coloured solid is formed which has a greater electrical conductivity than pure graphite.

These substances are referred to as graphitic compounds even though, strictly speaking, they are not compounds in the usual sense of the word.

Another example of a macromolecular layer lattice is a form of boron nitride. It has the structure shown below.

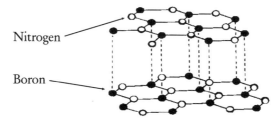

Nitrogen

Boron

Unlike graphite this substance is an insulator.

1　With reference to graphite, explain what you understand by the terms underlined in the passages.

2　Name the 'weak forces' which hold the layers together in the graphite structure.

3　Complete the electron configuration for the caesium atom:

Cs [Xe] .....

4　Explain with the aid of a diagram the bonding found in caesium metal.

5　Explain how you think the introduction of caesium into the graphite structure may have increased the electrical conductivity of the 'graphitic compound' so produced.

6　Suggest a reason why the boron nitride, shown in the diagram, is an insulator.

7　Graphite is used in some nuclear reactors. The rate of nuclear reactions in such reactors is controlled by inserting moderators, such as graphite, into the reactor core. Explain what you understand by the term moderator in this context.

8　Industrial diamonds are made from graphite by making use of the following equilibrium.

$$C \text{ (graphite)} \rightleftharpoons C \text{ (diamond)} \qquad \Delta H^{\ominus} = +2.1 \text{ kJ mol}^{-1}$$

Given that diamond is the denser of the two allotropes, suggest possible conditions of pressure and temperature which could be used to produce diamond from graphite.

13   Effusion

In an effusion experiment the apparatus was set up as shown. The oven surrounding the gas syringe ensured that the temperature was greater than the boiling point of either of the two volatile liquids ethoxyethane $(CH_3CH_2OCH_2CH_3)$ and unknown liquid X.

Using the apparatus shown it was found that, at constant temperature, 50 cm^3 of ethoxyethane vapour effused through the hole in 20 seconds and that 80 cm^3 of vapour X effused through the same hole in 35 seconds.

Gas syringe

Syringe oven →

Light bulb
heat source

Small hole

1 Calculate the relative molecular mass of X.

2 Given that X is an aliphatic ester, write down the structural formulae of the isomers of X.

3 A sample of the aliphatic ester X was treated with aqueous sodium hydroxide and potassium iodide, and a yellow precipitate was obtained. Using this information, write down the structural formula of X.

Natural uranium is found as a mixture of two isotopes, ^{235}U and ^{238}U. ^{238}U does not undergo fission when irradiated with neutrons and so it must be separated from ^{235}U which does undergo fission and is used in nuclear reactors. The two isotopes are separated by a process of effusion after the conversion of the uranium into gaseous UF_6. The UF_6, containing $^{235}UF_6$ and $^{238}UF_6$, is passed along a pipe made of a porous material which allows both gases to pass through. The gases are collected in an outer pipe. The $^{235}UF_6$ effuses from the porous pipe faster than the $^{238}UF_6$ and so the proportion of $^{235}UF_6$ to $^{238}UF_6$ in the outer collecting pipe increases. Therefore, the UF_6 initially produced is separated by enriching the $^{235}UF_6$. Several kilometres of piping are needed to give the required separation.

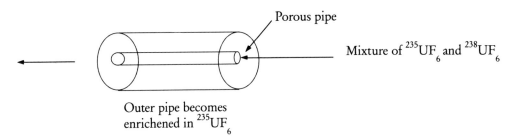

Porous pipe

Mixture of $^{235}UF_6$ and $^{238}UF_6$

Outer pipe becomes
enriched in $^{235}UF_6$

4 a State the law which applies to gaseous effusion.

  b Using this law, calculate the ratio of rates of effusion of the gaseous fluorides.

14 Relative Molecular Mass

The apparatus below is that which can be used to determine the relative molecular mass of a volatile liquid.

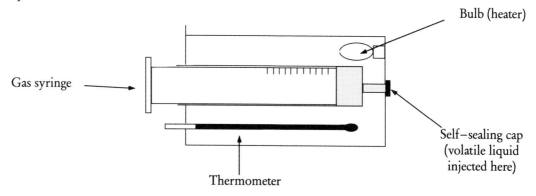

Since measurements are being made in the gaseous state at constant high temperature, it can be assumed that, the IDEAL GAS EQUATION may be applied, i.e.

$$pV = nRT$$

where:

p=pressure, V=volume, n=number of moles, R=gas constant and T=temperature.

Since
$$n = m/M$$

we can re-write the ideal gas equation in the form:

$$pV = \frac{mRT}{M}$$

1 Use this equation to calculate the molecular mass of volatile liquid X based on the information given below:

   0.178 g of liquid X was injected into a gas syringe maintained at 100 °C.

   This liquid produced 94 cm^3 of vapour at this temperature.

   The experiment was carried out at 1 atmosphere pressure.

2 X is an aliphatic ketone. Write a possible structural formula for X as well as the corresponding isomeric aldehyde.

3 a Treatment of X with NaBH$_4$ in aqueous solution gave a compound Y. Write the structural formula of Y and state the type of reaction taking place.

   b Write the structural formula of the compound which would be produced if the isomeric aldehyde was treated with NaBH$_4$ in aqueous solution.

4 Compound X can be identified by forming a crystalline derivative with 2,4–dinitrophenylhydrazine. The melting point of this derivative can be used to identify compound Y. Write a possible structural formula for this derivative and state the type of reaction that occurs when it is formed.

5 Describe a chemical test which would allow you to distinguish between X and the isomeric aldehyde.

15 Anti—Cancer Drug

Chlorambucil is a drug used in the treatment of leukaemia. It is one of the <u>slowest acting drugs</u> and is effective not only against some types of leukaemia but also against certain other cancers. Chlorambucil has the structure:

1 Identify two functional groups found in this compound.

2 Explain why the 'slow acting' nature of this drug would be an advantage.

3 Draw the structures of the expected organic products from the reaction of Chlorambucil with:
  a dilute potassium hydroxide,
  b ethanol (in the presence of concentrated sulphuric acid),
  c potassium cyanide in alcoholic solution.

4 a Outline the mechanism for the reaction which takes place in question **3c**.
  b What type of reaction is taking place ?

5 a Chlorambucil forms a solution with a pK_a of 4.9 at 25 °C. Calculate a value of K_a for this substance.
  b Given that the pK_a of benzoic acid is 4.2 at the same temperature (i.e. 25°C), state the main cause of the difference in acid strengths of these two substances.

Iron, which is the second most abundant metal in the earth's crust, is one of the most important of metals. The whole of our way of life is based on the use of machinery and consumable goods made of iron or alloys containing iron.

The extraction of iron is carried out in a blast furnace. Blast furnaces vary in height from 25–60 m and the width can be up to 10 m.

Read through the sequence below which outlines the reactions used in extracting iron from its ore haematite in the blast furnace.

$$C_{(s)} + O_{2(g)} \longrightarrow X_{(g)} \qquad \Delta H_1$$

$$X_{(g)} + C_{(s)} \longrightarrow Z_{(g)} \qquad \Delta H_2$$

$$\text{Haematite} + Z_{(g)} \longrightarrow Fe_{(l)} + X_{(g)} \quad \Delta H_3$$

1 Identify each of the materials X and Z by name and formula. Complete and balance the chemical equations for the three reactions of the sequence.

2 Which of the above enthalpy changes is mainly responsible for the high temperature developed in the body of the furnace ?

Limestone is added to remove the sandy impurities from the ore to form a slag. The sequence below outlines the reactions involved.

$$\text{Limestone} \longrightarrow A_{(s)} + X_{(g)}$$

$$A_{(s)} + SiO_{2(s)} \longrightarrow \text{calcium silicate (slag)}$$

3 Identify the material A by name and formula. Complete and balance the equations for the two reactions of the sequence.

0.20 g of impure iron (containing iron and carbon only) was dissolved in an excess of dilute sulphuric acid. The resulting pale green solution G was then titrated with potassium manganate(VII). Solution G decolourised 25.00 cm^3 of 0.02M potassium manganate(VII).

4 Write down the formula of the iron–containing species in solution G.

5 Describe what you would see if dilute aqueous ammonia was added dropwise, to excess, to solution G and the resulting mixture allowed to stand.

6 Explain the observations you have described in question **5**.

7 Write the balanced equation for the reaction of $Fe^{2+}_{(aq)}$ ions with $MnO_4^{-}_{(aq)}$ ions in acidified aqueous solution.

8 Calculate the percentage of iron in the impure sample.

Iron has an essential biological role. The daily requirement for humans is 0.01 mg. It is needed for the production of haemoglobin, the red pigment found in blood.

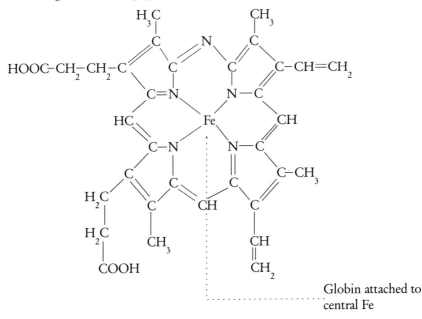

The central metal ion is the iron(II) ion.

9   What do you understand by the term ligand ? The haem ligand is a 'tetradentate' ligand. With reference to the diagram above, explain the meaning of this term.

10   **a**   Name the substance carried to the body tissues by haemoglobin.

     b   State the number of molecules of the substance you have named in question **10 a** , which are carried by each molecule of haemoglobin.

11   Which element in the haem ligand is bonded to the iron(II) ion, and what sort of chemical bond is involved in this link ?

12   Name an illness caused by a deficiency of iron.

18   © Earl and Wilford 1992

17   An Analgesic

An analgesic, or painkiller, is a substance which relieves pain. Paracetamol is a commonly used analgesic. It has the following structure:

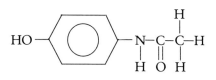

1 Identify two functional groups in this compound.

2 Suggest a possible problem associated with the preparation of paracetamol starting from 4–aminophenol and ethanoyl chloride.

3 Draw the structures of the expected organic products from the reaction of paracetemol with:

    a chlorine in the presence of ultra–violet light,

    b phosphorus(V) chloride,

    c dilute hydrochloric acid.

4   **a** Outline the mechanism of the reaction taking place in question **3 a**.

    b What type of reaction is this ?

5 Give a simple chemical test which would allow you to identify the presence of the CH_3CO group in this compound.

6 Paracetamol forms a weakly acidic solution which has a pK_a of 9.5 at 20 °C.

    a Define pK_a.

    b Calculate a value of K_a for paracetamol at 20 °C.

    c Account for the acidic character of paracetamol.

18 Titanium

Titanium is the second member of the first transition series of metals. It is the seventh most abundant metal occuring in the earth's crust (0.6%) and is a metal of the same kind of mechanical strength as steel. However, it has two main advantages over steel:

a) it does not corrode,

b) it is less dense (about 60% of that of steel).

It is not possible to obtain titanium by the common method of reduction with carbon because a very stable carbide is produced. Also, it reacts with oxygen and nitrogen at high temperatures. It is these difficulties which delayed the commercial production of the metal until about 1950.

Read through the sequence below which outlines the reactions used in extracting titanium from rutile, i.e. titanium(IV) oxide.

$$\text{Rutile} \ + \ \text{carbon} \ + \ R_{(g)} \longrightarrow X_{(l)} \ + \ CO_{2(g)}$$

$$X_{(l)} \ + \ Mg_{(s)} \longrightarrow Ti_{(s)} \ + \ Z_{(s)}$$

1 Identify the materials R, X and Z. Complete and balance the chemical equations for the two reactions.

2 Name an alternative metal which can be used instead of magnesium to release titanium metal from X.

3 How is product X of the first reaction obtained in a pure form to use in the second stage of the process ?

4 The second stage of the process involves heating X with magnesium in an atmosphere of argon. Why is this precaution necessary ?

5 Give two major uses of titanium.

6 Even though titanium is corrosion resistant in the powdered form it will burn very readily in oxygen as well as liberating hydrogen from steam in a similar manner to other metals. Write balanced chemical equations to represent both of these reactions.

7 a Titanium is classified as a 'transition element' whilst zinc is not. Explain this statement.

   b Complete the electronic configuration of:

   i Ti   [Ar] ....

   ii Ti^{2+} [Ar] ....

   iii Ti^{4+} [Ar] ....

8 A is a white powder. It is a compound of zinc which when heated produces a colourless gas D and leaves a yellow residue E which turns white on cooling. E dissolves in dilute sodium hydroxide producing a colourless solution G. E also reacts with dilute sulphuric acid producing a colourless solution J.

   a Identify A, D and E by name and formula.

   b Write down the formula of the zinc–containing species in solution G.

   c Write a balanced chemical equation for the reaction of E with dilute sulphuric acid.

   d Write down the formula of the zinc–containing species in solution J.

   e Substance E is said to be amphoteric. Using the information given in the above short passage, explain fully what you understand by this term.

19 Reactions of Anaesthetics

An anaesthetic is a substance which has the ability to abolish pain and other sensations (such as touch and temperature). Novocaine is a local anaesthetic which is often used for minor surgery to areas such as the nose, throat or eyes. It is applied to the surface before surgery. Novocaine has the following structure:

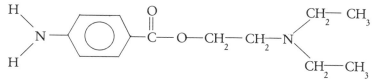

1 Name two functional groups found in this substance.

2 Novocaine reacts with nitrous acid at 5 °C.

   a Draw the structure of the expected organic product of this reaction.

   b Name the type of reaction which has taken place and note any other possible organic products associated with this reaction.

   c When phenol in alkaline solution is added to the organic product of this reaction, at 5 °C, a dye is produced. Draw the structure of the dye molecule and note any other possible organic products associated with this reaction.

3 Novocaine is slightly basic with a K_b value of 4.1×10^{-7} mol dm^{-3}.

   a Account for this basic character in terms of its structure.

   b Calculate a value of pK_b for Novocaine.

In 1956, ICI chemists produced a haloalkane which has wide spread use for major surgery in hospitals. This substance was marketed as 'Fluothane' (or 'Halothane') and has the structure:

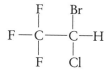

4 What is the systematic name for Fluothane ?

5 Haloalkanes undergo 'nucleophilic substitution' reactions. Explain the meaning of this term.

6 Mean carbon–halogen bond enthalpies are shown in the table below.

C – Halogen Bond	Mean Standard Bond Enthalpy/kJ mol^{-1}
C–Cl	338
C–I	238
C–F	484
C–Br	276

   a Which of the carbon–halogen bonds is the strongest ?

   b Give the order of ease of bond breaking which will take place in a typical nucleophilic substitution reaction involving a haloalkane.

   c Under normal laboratory conditions, Fluothane will react with aqueous sodium hydroxide. By taking into consideration your answers to parts a and b, draw the likely structure of the product of this reaction.

20 Tackling Sleeping Sickness

The drug difluoromethylornithine (DFMO):

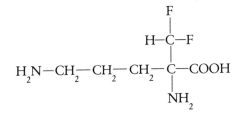

was originally developed by researchers at the University of California, San Fransisco, as a cancer treatment. However, it is now proving to be effective against 'sleeping sickness' and is the first new drug to help combat this disease in nearly 50 years.

1   In order to identify the $-NH_2$ functional group in DFMO it was treated with nitrous acid followed by an alkaline solution of phenol.

   a   Suggest the reagents and conditions used to produce nitrous acid.

   b   What conditions would be required for the reaction between DFMO and nitrous acid ?

   c   What would you observe during the reaction in part **b** ?

   d   Write an equation for the reaction of DFMO with nitrous acid.

   e   How may the rate of this reaction be followed ?

2   What feature of the molecule of DFMO accounts for its optical activity ?

3   Identify two other functional groups found in this compound other than that responsible for the reactions referred to in question **1**

4   Draw the structures of the expected organic products for the reaction of DFMO with:

   a   ethanol in the presence of concentrated sulphuric acid,

   b   phosphorus(V) chloride,

   c   lithium tetrahydridoaluminate dissolved in ethoxyethane.

5   The purity of individual compounds used to treat disease is very important. A 1.88 g sample of freshly prepared DFMO was reacted with sodium hydroxide. After separation, 1.88 g of a compound containing sodium was obtained. Calculate the percentage purity of this sample of DFMO.

21 Enthalpy of Combustion

An experiment was carried out to determine the enthalpy of combustion of butan–1–ol. The apparatus used is shown opposite. The spirit burner, with butan–1–ol, was first weighed and the value recorded. It was allowed to burn for five minutes under a copper calorimeter containing 500 cm^3 of water whose initial temperature was also recorded.

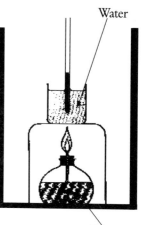

Water

Butan–1–ol

After five minutes the mass of the spirit burner was recorded again as was the temperature of the water.

The results are shown below.

Initial mass of the spirit burner + butan–1–ol　　= 29.42 g

Final mass of the spirit burner + butan–1–ol　　= 27.51 g

Initial temperature of water in the copper calorimeter　= 21.2 °C

Final temperature of water in the copper calorimeter　= 44.1 °C

1　Assuming the specific heat capacity of water is 4.2 J g^{-1} K^{-1}, its density is 1 g cm^{-3} and that the specific heat capacity of the copper calorimeter is negligible, calculate the enthalpy change during the experiment.

2　Calculate the enthalpy change per mole of butan–1–ol from this experiment. (R.A.M., A_r C=12, H=1, O=16)

3　**a**　A data book value for the standard enthalpy of combustion of butan–1–ol is –2670 kJ mol^{-1}. Calculate the percentage uncertainty in the experimental value with that of the data book value. Explain the possible sources of uncertainty.

　　b　Given that the standard enthalpies of combustion of methanol, ethanol and propan–1–ol are –726, –1370 and –2010 kJ mol^{-1} respectively, plot a graph of enthalpy of combustion against relative molecular mass (RMM) for these three alcohols and for butan–1–ol.

4　From your graph in question **3**, predict the molar enthalpy of combustion of pentan–1–ol. What assumption(s) have you made in making this prediction ?

5　What can you say about the energy produced when alcohols of progressively higher RMM are burned in air?

22   Reaction Kinetics

Experiments were carried out to study the kinetics of the reaction:

$$H_2O_{2(aq)} + 2H^+_{(aq)} + 2I^-_{(aq)} \longrightarrow 2H_2O_{(l)} + I_{2(aq)}$$

The rate of the reaction was measured by taking the time for sufficient iodine to be formed and to react with a fixed amount of sodium thiosulphate. Any excess iodine was detected by the formation of a blue colour with starch solution.

A stop clock was started when 2 cm^3 of sodium thiosulphate solution and 2 cm^3 of starch solution were added to different volumes of 0.1M hydrogen peroxide solution, 0.5M sulphuric acid solution and 0.1M potassium iodide solution, distilled water being added each time to maintain a constant volume. The clock was stopped when the blue colour appeared.

The results obtained are shown below.

Experiment	Vol. 0.1M KI/cm^3	Vol. 0.5M H$_2$SO$_4$/cm^3	Vol. 0.1M H$_2$O$_2$/cm^3	Vol. H$_2$O/cm^3	Total vol./cm^3	Time/s
1	6	6	6	6	24	56
2	4	6	6	8	24	85
3	2	6	6	10	24	175
4	6	4	6	8	24	56
5	6	2	6	10	24	56
6	6	6	4	8	24	87
7	6	6	2	10	24	179

1 Fill in the table below by:

    a calculating the initial rate of the reaction for each experiment. This rate is inversely proportional to the time taken for the blue colour to appear,

    b calculating the number of moles per dm^3 of the reagents in each experiment.

Experiment	[I$^-$]	[H$^+$]	[H$_2$O$_2$]	Rate 1/t sec^{-1}
1				
2				
3				
4				
5				
6				
7				

3 Plot three graphs, each showing how varying the concentration of each of the reactants in turn, whilst the others are kept constant, affects the rate of reaction.

4 On which reactant concentration(s) does the reaction rate depend ?

5 Deduce the order of reaction with respect to:

    a [I$^-$],   **b** [H$^+$], and   **c** [H$_2$O$_2$]

    Hence write a full rate equation for the reaction.

6 What is the overall order of this reaction ?

7 Suggest a possible mechanism for this reaction.

23 Newly Discovered Elements

The International Union of Pure and Applied Chemistry (IUPAC) is the world authority on the names of chemical elements and their compounds. To date, IUPAC have recognised, and have named, the elements up to 107 which has been named Unnilseptium and given the symbol Uns. Symbols, mass numbers, as well as relative atomic masses and half–lives are known for the newly discovered elements 104 to 107 and some are shown in the table below.

Symbol	Atomic Number (Z)	Mass Number (A)	Relative Atomic Mass	Half–Life /s
Unq	104	261	261.11	65
Unp	105	262	262.114	–
Unh	106	263	263.118	0.9
Uns	107	264	262.12	0.12

1 What do you understand by the term 'half–life' of a radioactive element ?

2 The radioactivity of the element unnilpentium (Unp) was monitored over a period of four minutes using a Geiger–Müller tube. The results of this exercise are shown in the table below.

Time/s	0	30	60	90	120	150	180	210	240
Radioactivity/counts	500	265	150	77	45	27	15	6	4

Use the data above to plot a graph which will enable $t_{1/2}$ of Unp to be calculated.

3 Using the information above, calculate the mass of ^{261}Unq left after 0.004 g of this substance has decayed over a period of 195 seconds.

4 a State the relative masses and relative charges of the two particles which are generally emitted by radioactive elements.

   b ^{263}Unh is thought to undergo alpha decay to give the radioactive element X. X is thought to undergo beta decay followed by further alpha decay to the elements Y and Z respectively. Write the equations for the nuclear reactions, using your Periodic Table as well as the table above to identify the elements X, Y and Z.

5 How many neutrons are there in each atom of ^{263}Unh and ^{264}Unp ?

6 Why are the relative atomic masses shown in the table not whole numbers ?

© Earl and Wilford 1992

24 Sulphuric Acid

Sulphuric acid is a colourless, viscous liquid. It is used on an enormous scale; in the United Kingdom alone, 2.5 million tonnes of the concentrated acid are manufactured and used in the production of:

detergents, fertilisers, dyestuffs, explosives, paints and many other substances.

Study the reaction scheme below. This highlights the manufacture as well as some reactions of this useful inorganic acid.

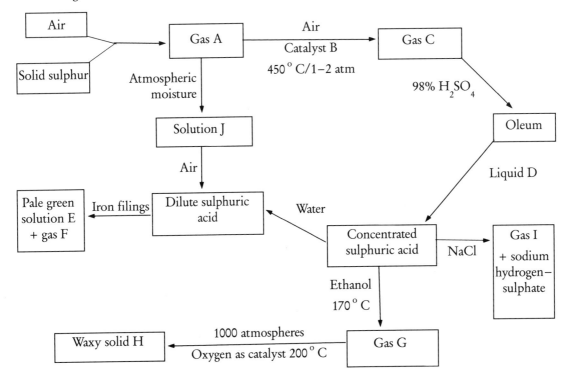

1 Identify the substances A – J giving their names and formulae.

2 a Write a balanced chemical equation for the formation of gas C.

 b Discuss the choice of conditions used to bring about the reaction you have written in part **a**, given that the formation of C is an exothermic process.

3 What is the role of concentrated sulphuric acid in its reaction with ethanol in the reaction scheme?

4 a What is the general name given to the process by which gas G is converted into the waxy solid H?

 b By what reaction mechanism is the waxy solid H formed?

5 Give the formula of the iron–containing species formed when iron filings are added to dilute sulphuric acid.

6 Describe a chemical test which could be used to identify gas G.

7 a Describe how the reactions of concentrated sulphuric acid with a separate solid bromide, and iodide, can be used to identify the halide ion present in these solids.

 b What is the role of the concentrated sulphuric acid in its reaction with the substances in part **a**?

8 a Gas A is one of the chief culprits in the formation of 'acid rain'. Natural sources of gas A are plankton, rotting vegetation and volcanoes. However, over half of this gas comes from the combustion of fossil fuels. When gas A passes into the atmosphere it reacts with moisture forming acid solution J. Write chemical equations for the formation of J and its subsequent oxidation to sulphuric acid.

 b Describe one measure that is being taken to help solve the problem of 'acid rain'.

Ammonia is a colourless gas with a choking smell. It is less dense than air. It is usually made in the laboratory by heating an ammonium compound with an alkali. The usual substances used are ammonium chloride and calcium hydroxide.

1 Write a balanced chemical equation for the production of ammonia using the substances named above.

The ammonia gas is dried by passing it through a drying tower.

2 Which substance could be used to dry the gas ?

Industrially ammonia is manufactured by the Haber Process. This process was initially developed by a German scientist called Fritz Haber, in 1911. By 1913, his process was further improved by Bosch to make ammonia in large quantities, to satisfy:

a) the need for Germany to produce large quantities of explosives,

b) the much increased demand for artificial fertilisers.

Today 85% of the ammonia produced is used in making fertilisers. Nitram$^{©}$ is a nitrate fertiliser manufactured by ICI. It is manufactured by the reaction of nitric acid with ammonia solution according to the equation:

$$NH_{3(aq)} + HNO_{3(aq)} \longrightarrow NH_4NO_{3(aq)}$$

3 A bag of Nitram$^{©}$ may contain 50 kg of ammonium nitrate. What mass of nitric acid would be required to make this amount of ammonium nitrate ?

In the above reaction ammonia is acting as a base.

4 What do you understand by the terms:
  a Brönsted–Lowry base,
  b Lewis base ?

Ammonium nitrate is the most common nitrate fertiliser used today. This means that large amounts of nitric acid have to be made in addition to the ammonia. Modern fertiliser factories contain a nitric acid plant to do this as well as an ammonia plant. Nitric acid is made by the oxidation of ammonia in the presence of a catalyst.

The concentration of an acid such as nitric acid may be determined by the technique of TITRATION. An acid–base titration is a method by which acid is added quantitatively to a base until the solution produced contains a salt and water only.

The following results were obtained for the neutralisation reaction of 25 cm^3 of dilute sodium hydroxide with dilute nitric acid. The initial concentration of the dilute sodium hydroxide was 0.3M.

The solution in the burette was dilute nitric acid.

The indicator used was phenolphthalein.

Titration results.

	Approximate	1	2	3
Final burette reading/cm^3	20.50	26.40	27.80	25.30
Initial burette reading/cm^3	0.00	6.20	7.60	5.00
Volume used/cm^3				

5   From the specimen results, calculate the volume of dilute nitric acid used in each titration.

6   From the three most accurate results, calculate the average volume of dilute nitric acid used.

7   a Write a balanced chemical equation for the reaction which has taken place during the titration.

   b Write down the number of moles of nitric acid and sodium hydroxide shown reacting in the equation.

8   a Calculate the number of moles of sodium hydroxide present in 25 cm^3 of solution.

   b Calculate the number of moles of nitric acid neutralised.

9   Calculate the molarity of the dilute nitric acid.

10 Identify further important uses of nitric acid besides the manufacture of fertilisers.

26   Chloride Analysis

The formula of any soluble ionic chloride can be determined by a precipitation reaction using silver nitrate solution. The reaction can be carried out in the form of a titration using aqueous potassium chromate(VI) or fluorscein as the indicator.

The following results were obtained for such an experiment:

0.85 g of an anhydrous Group 2 chloride was dissolved in distilled water. This solution was made up to 250 cm^3 in a volumetric flask. 25 cm^3 of the resulting solution required exactly 30 cm^3 of 0.05M silver nitrate to be added before the end–point of the titration was observed.

1   Describe the colour change which takes place at the end–point of this titration.

2   Write the equation for the reaction which takes place between silver ions and chloride ions.

3   Write down the number of moles of chloride ions which react with one mole of silver ions.

4   Calculate the number of moles of silver ions used in this titration.

5   Calculate the number of moles of chloride ions reacting with the number of moles of silver ions which you have calculated in question **4**.

6   Calculate the mass of chloride ions from your answer to question **5**.
    (R.A.M., A_r Cl=35.5)

7   Calculate the mass of chloride ions present in 250 cm^3 of the Group 2 metal chloride solution.

8   Calculate the mass of Group 2 element present in 250 cm^3 of solution.

9   The relative atomic masses of the Group 2 metal element and chlorine are 40 and 35.5 repectively. Use this information to determine the empirical formula for this chloride.

10 Comment on any experimental uncertainties which you think may contribute to an empirical formula determination using this technique.

11 Silver chloride will dissolve in aqueous ammonia to produce a colourless solution.
   a Name and give the formula of the complex ion of silver present in the resulting colourless solution.
   b Draw the structure of the complex ion you have named in part **a**, giving the relevant bond angle(s).

12 Describe, with full experimental details, how it might be possible to compare the amount of common salt present in two different brands of 'ready salted' crisps, using the technique outlined above.

27   Household Bleaches

Bleaches such as Domestos and Parazone contain sodium chlorate(I), (also known as sodium hypochlorite). This substance can be manufactured by the reaction of chlorine with sodium hydroxide solution according to the following ionic equation:

$$\text{Chlorine}_{(g)} + A_{(aq)} \longrightarrow B_{(aq)} + ClO^-_{(aq)} + H_2O_{(l)}$$

1 Identify each of the ions A and B by name and formula.

2 Complete and balance the ionic equation.

In acid solution, chlorate(I) ions react releasing chlorine as follows:

$$OCl^-_{(aq)} + Cl^-_{(aq)} + 2H^+_{(aq)} \longrightarrow H_2O_{(l)} + Cl_{2(g)}$$

The chlorine released is referred to as 'available' chlorine. The available chlorine so produced can be estimated as a percentage (w/v) by:

a) initial displacement of iodine from potassium iodide,

b) titration of the displaced iodine using standard sodium thiosulphate.

The following data were obtained from an experiment of this nature:

10 cm^3 of bleach were made up to 250 cm^3 with distilled water in a volumetric flask. 25 cm^3 of this solution were pipetted into a conical flask. 15 cm^3 of a 10% solution of potassium iodide were added followed by 15 cm^3 of dilute ethanoic acid. The liberated iodine was titrated against standard 0.1M sodium thiosulphate using starch as an indicator. The blue colour of the starch disappeared upon addition of 24.17 cm^3 of this standard solution.

3 Write down the ionic equation which represents the reaction of the 'available' chlorine with iodide ions.

4 Write down the equation which represents the reaction of the liberated iodine with thiosulphate ions.

5 From the information in the passage above, determine the number of moles of chlorine liberated.

6 Calculate the molarity of the chlorine solution using the data you have obtained in question **5**.

7 Calculate the concentration of the diluted bleach in $g \text{ dm}^{-3}$.

8 Using your answer to question **5**, calculate the percentage of 'available' chlorine in the bleach solution.

9 Ethanoic acid is added to release the chlorine from the bleach. Why cannot dilute hydrochloric acid be used for this purpose ?

10 What is the oxidation state of chlorine in OCl^- ?

11 'Environmentally friendly' bleaches do not contain sodium chlorate(I) (sodium hypochlorite). They contain, instead, hydrogen peroxide.

What is the chemical advantage of using bleaches which contain hydrogen peroxide over those containing sodium chlorate(I) ?

Aspirin was first made in 1899 and therefore is one of the earliest drugs. It is a drug which lowers the body temperature rapidly and effectively in feverish patients (an antipyretic) as well as a pain killer (an analgesic). The main constituent of commercial asprin tablets is 2–ethanoylhydroxybenzoic acid. The route to its manufacture can be seen in the following reaction scheme starting with phenol.

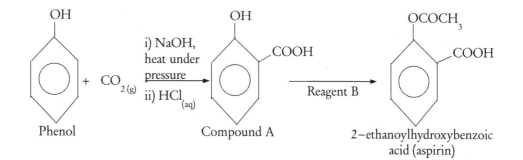

However, aspirin can cause some stomach bleeding and some people need hospital treatment after taking only small quantities of the drug.

1   Name compound A and identify by name and formula a possible reagent B.

2   Write a balanced chemical equation to represent the reaction of compound A with reagent B.

3   Name an alternative analgesic to aspirin which does not cause stomach bleeding.

4   A functional group found in the analgesic you have named in question **3** is CH_3CO-. A functional group found in aspirin is $-COOH$. Describe simple chemical tests which may be used to identify the presence of these functional groups within an organic compound.

5   2–ethanoylhydroxybenzoic acid may contain an intra–molecular hydrogen bond. Draw a diagram to show the structure of 2–ethanoylhydroxybenzoic acid containing this bond.

6   Draw the structures of the main organic products for the reaction of aspirin with aqueous sodium hydroxide.

7   In which part of the digestive system does a reaction similar to that in question **6** take place ?

The following experiment was carried out on aspirin tablets:

1.5 g of aspirin were placed in a clean conical flask. This is equivalent to four tablets. 10 cm^3 of a 0.01M sodium hydroxide solution were pipetted onto the powdered aspirin in a conical flask. The mixture was then warmed and stirred for approximately ten minutes. The mixture was then cooled and made up to 250 cm^3 with distilled water.

25 cm^3 of this solution were then titrated against 0.1M hydrochloric acid using phenolphthalein as the indicator. This solution required 25.5 cm^3 of dilute acid for neutralisation.

8 Write a balanced chemical equation for the reaction which took place upon the addition of sodium hydroxide to the aspirin tablets.

9 Calculate the number of moles of sodium hydroxide which were added to the powdered aspirin tablets in the conical flask.

10 Write a balanced chemical equation for the reaction which takes place during the titration of the excess sodium hydroxide.

11 Calculate the number of moles of sodium hydroxide which were in excess.

12 How many moles of 2−ethanoylhydroxybenzoic acid have reacted with the sodium hydroxide solution ?

13 Calculate the mass of 2−ethanoylhydroxybenzoic acid present in the aspirin tablets used.

(R.A.M., A_r C=12, H=1, O=16)

14 What is the percentage of 2−ethanoylhydroxybenzoic acid found in the powdered aspirin ?

15 Suggest a bulking agent which might be used to make up the remaining mass of these aspirin tablets.

29   A Useful Salt

Potassium chloride is used as a fertiliser to counteract the deficiency of potassium in the soil. It is also used as a substitute for common salt.

Potassium chloride is formed from potassium (atomic number 19) and chlorine (atomic number 17). The lattice energy of this substance may be calculated by using a Born–Haber Cycle from the following experimental data:

	kJ mol^{-1}
First ionisation energy of potassium =	+419
Enthalpy of formation of potassium chloride =	−436.7
Enthalpy of atomisation of potassium =	+89.2
Electron affinity of chlorine =	−349
Enthalpy of atomisation of chlorine =	+122

1   Write an equation in each case to define the terms shown in the table above.

2   Construct a labelled Born–Haber Cycle for the formation of potassium chloride.

3   From your constructed Born–Haber Cycle, calculate the lattice energy of potassium chloride.

4   **a**   Complete the electron configurations for the following atoms and ions:

     i K   [Ne]. . .

     ii Cl   [Ne]. . .

     iii K$^+$ [Ne]. . .

     iv Cl$^-$ [Ne]. . .

   b   Draw a diagram which shows the type of bonding present in solid potassium chloride.

5   Aqueous potassium chloride reacts with silver nitrate solution whilst solid potassium chloride reacts with concentrated sulphuric acid.

   a   What observations would you expect in both of these reactions?

   b   Write a balanced equation in each case to represent these observations for the two reactions taking place.

6   Potassium chloride occurs naturally as sylvite as well as in a hydrated double salt with magnesium chloride, carnallite.

   a   Explain the meaning of the terms 'double salt' and 'hydrate'.

   b   The following percentage composition by mass was obtained by analysis of a sample of carnallite from the Stassfurt salt deposits.

Element	Percentage Composition by Mass
K	14.05
Cl	38.40
Mg	8.65
H	4.32
O	34.60

Using this data and given that the M_r of the double salt is 277.5, calculate the molecular formula of this sample of carnallite.

(R.A.M., A_r K=39, Cl=35.5, Mg=24, H=1, O=16).

30   A Versatile Transition Metal

A transition or d–block element is defined as an element in which there is an incomplete subshell of d–electrons. These elements show variable oxidation states and produce coloured compounds. The first transition series occurs between Groups 2 and 3 of the Periodic Table. This series consists of ten metals, scandium to zinc. Zinc, however, has a full d–subshell and so strictly is not a transition metal. It is, however, usually included in the series. Transition metals and their compounds are important catalysts in industry and in biological systems as well as being important in the production of alloys.

An atom of the transition metal M has the electron configuration:

$$1s^2, 2s^2\, 2p^6, 3s^2\, 3p^6\, 3d^9, 4s^2$$

1   What is the identity of metal M ?

2   M can show two stable oxidation states, M^+ and M^{2+}, in its compounds. Give the electron configuration of these two oxidation states.

3   Give the formula of an oxide of M in its higher oxidation state.

4   When the metal M oxide reacts with dilute sulphuric acid it produces a blue solution.

   a   Give the formula of the complex ion present in this solution which gives rise to the blue colour.

   b   Show, by means of a sketch, how the water molecules are linked to the metal ion.

   c   What is the shape of this complex ion ? Explain, using the electron pair repulsion theory, why this shape occurs.

   d   What is the term used to describe the role of the water molecules in this complex ion ?

   e   Name this complex ion.

5   The metal(I) iodide is produced when potassium iodide solution is added to the blue solution in question **4**.

   a   Describe and account for the changes which take place when potassium iodide is added to the blue solution.

   b   Write an equation for the reaction taking place.

The standard electrode potentials of M^+ and M^{2+} are:

$$M^+_{(aq)} + e^- \rightleftharpoons M_{(s)} \qquad E^{\ominus} = +0.52\ V$$

$$M^{2+}_{(aq)} + e^- \rightleftharpoons M^+_{(aq)} \qquad E^{\ominus} = +0.15\ V.$$

6   **a**   Calculate the $E^{\ominus}$ for the equilibrium:

$$2M^+_{(aq)} \rightleftharpoons M_{(s)} + M^{2+}_{(aq)}$$

   b   What type of reaction is taking place in question **6 a** ?

   c   Comment on the stability of the M^+ ion in aqueous solution from the data above.

31   Natural Gas

Natural gas consists mainly of methane with a small proportion of other volatile hydrocarbons. However, it also contains a small proportion of carbon dioxide.

Carbon dioxide is removed from methane for a variety of reasons including:

a) minimisation of pipeline corrosion,

b) provision of acceptable burning characteristics,

c) liquefaction for storage and sale by the cylinder or tanker,

d) use as a chemical feedstock for ammonia synthesis.

British Gas removes carbon dioxide from natural gas (methane) by passing it through hot potassium hydroxide solution.

1   Write a balanced chemical equation for the reaction which takes place when excess carbon dioxide is bubbled through potassium hydroxide solution.

2   Explain how the removal of carbon dioxide from methane will reduce the amount of corrosion as the natural gas passes through steel pipes.

3   **a**   Write a balanced chemical equation for the complete combustion of methane.

      b   Explain why the presence of carbon dioxide tends to give unacceptable burning characteristics.

4   Explain why it is necessary to remove carbon dioxide before methane is liquified and stored in cylinders.

5   Why must carbon dioxide be removed before methane is used as a chemical feedstock for ammonia synthesis ?

Sulphur also needs to be removed from the natural gas because its presence would poison the catalysts used in the production of ammonia. The sulphur in natural gas is present as hydrogen sulphide which can be removed by reaction with zinc oxide.

6   Write a balanced chemical equation for the reaction between hydrogen sulphide and zinc oxide.

After the purification of natural gas, methane, is used as a major source of hydrogen for the Haber synthesis of ammonia. The hydrogen is produced by an equilibrium process involving 'steam reforming'. In this process methane is reacted with steam over a nickel catalyst at 750°C.

7   Write a chemical equation for the steam reforming process.

8   What is the role of the methane in the reaction with steam?

9   Suggest two possible ways of increasing the concentration of hydrogen in the equilibrium mixture at a given temperature. Explain your answer.

32 Internal Combustion Engine

The most important reaction of the alkanes is that of combustion. Generally, when alkanes burn they form relatively harmless products, water and carbon dioxide, via exothermic reactions. This type of reaction is taken advantage of in the internal combustion engine. The liquid alkane, octane, is vaporised in the engine and mixed with air. In the presence of a 'spark' from the spark plug the following complete combustion reaction takes place:

$$2C_8H_{18(g)} + 25O_{2(g)} \longrightarrow 16CO_{2(g)} + 18H_2O_{(g)}$$

1 Using the standard heats of formation data given in the table below, obtain a value for the $\Delta H_C^{\ominus}$ for this combustion reaction.

	$\Delta H_f^{\ominus}/\text{kJ mol}^{-1}$
$C_8H_{18(g)}$	−250.0
$CO_{2(g)}$	−394.0
$H_2O_{(g)}$	−242.0
$CO_{(g)}$	−111.0

2 If the supply of oxygen is insufficient, then incomplete combustion takes place and the poisonous gas, carbon monoxide, will be formed.

   a Write a balanced chemical equation for the incomplete combustion of octane.

   b Calculate a value for the enthalpy change for this incomplete combustion reaction.

   c Why are the answers you have obtained to questions 1 and 2 b different?

   d Why is the presence of carbon monoxide in exhaust gases a danger?

3 The extent of combustion of petrol vapour in an internal combustion engine can be shown in terms of the percentage of carbon dioxide found in the exhaust gases.

   120 cm³ of exhaust gases (assumed to contain carbon monoxide, carbon dioxide and nitrogen only) were mixed with 40 cm³ of oxygen and the resulting mixture exploded. After cooling, it was found that there had been a decrease in volume to 140 cm³. When potassium hydroxide solution was added the volume further decreased to 70 cm³.

   a Write a balanced chemical equation for the complete combustion of carbon monoxide.

   b Calculate the composition by volume of the mixture found in the exhaust gas.

   c Calculate the percentage by volume of carbon dioxide in the exhaust gases.

   d Write a balanced chemical equation for the reaction of potassium hydroxide with an excess of carbon dioxide.

33   Health Drink

Many health drinks are marketed today with claims that they combat colds and 'flu'. Many of these contain the analgesic paracetamol. Analgesics are substances that relieve pain. The term 'pain killers' has been used to describe this effect. A typical 'lemon' health drink has the following ingredients listed on the packet:

> Each 5.0 g packet contains the following active ingredients shown as percentage by mass:
>
> 60 % Glucose
>
> 1 % Ascorbic acid
>
> 0.5 % Phenylephrine hydrochloride
>
> 10 % Paracetamol
>
> 15 % Sodium citrate

1   Calculate the mass of each of the active ingredients in a 5.0 g packet of the health drink.

2   Given that the chemical formula of glucose is $C_6H_{12}O_6$ and that of paracetamol is $C_8H_9O_2N$, calculate the number of moles of each present in a 5.0 g packet of the health drink.

3   Sodium citrate is formed by the neutralisation of citric acid by sodium hydroxide solution. Citric acid is a tricarboxylic acid which has the structure:

$$\begin{array}{c} OH \\ | \\ HOOC-CH_2-C-CH_2-COOH \\ | \\ COOH \end{array}$$

     a   Write a balanced chemical equation for the complete neutralisation of citric acid.

     b   Name a suitable indicator for use with this titration.

     c   Calculate the mass of sodium citrate produced when just sufficient citric acid is added to 30 cm^3 of 1M sodium hydroxide to react completely.

     d   Write an overall expression for K_a for citric acid in aqueous solution.

4   Suggest reasons why glucose and ascorbic acid are included as ingredients in this health drink.

34   Chemical Rockets

Chemical rockets are used today in quite large numbers to put satellites into orbit. One version of an advanced liquid propellant system uses dinitrogen tetroxide (N_2O_4) as the oxidant with the hydrazine derivative, methylhydrazine, as the fuel. The resulting reaction is highly exothermic and nitrogen, carbon dioxide and water are produced according to the following equation:

$$5N_2O_{4(g)} + 4CH_3NHNH_{2(g)} \longrightarrow 9N_{2(g)} + 4CO_{2(g)} + 12H_2O_{(g)}$$

1   What do you understand by the term 'oxidant' as applied to this system?

2   A chemical rocket burns 200 dm^3 of methylhydrazine, as a gas, per second. Calculate :

   a   the volume of N_2O_4 used per second for complete combustion,

   b   the volume of nitrogen, carbon dioxide and water produced per second.

3   An excess of N_2O_4 is usually carried on a rocket. Why is it necessary to do this?

Dinitrogen tetroxide is a colourless gas at room temperature and pressure. At room temperature it is involved in the following equilibrium reaction:

$$N_2O_{4(g)} \rightleftharpoons 2NO_{2(g)} \qquad \Delta H = +ve$$

Nitrogen(IV) oxide is a brown gas at room temperature and pressure.

4   Describe what would happen to the equilibrium if the temperature was increased. Explain your answer.

5   Describe what would happen to the equilibrium if the pressure was increased. Explain your answer.

6   Nitrogen(IV) oxide is described as a pollutant. Discuss briefly your understanding of this statement.

35　　North Sea Analysis

Sea water contains many dissolved solids. For example, a 1 kg sample of water taken from the North Sea contains about 35 g of dissolved solid material. An analysis of these solids shows that the salts present in the sample are:

Salt	Mass present/g
Calcium carbonate	0.14
Calcium sulphate	1.37
Magnesium chloride	2.73
Magnesium sulphate	2.61
Potassium bromide	0.08
Potassium chloride	0.70
Sodium chloride	27.40

1　Which of the salts shown in the table is used:

　a　for medical purposes as a purgative,

　b　to cure soil acidity and in the manufacture of glass,

　c　as a salt substitute,

　d　as a sedative,

　e　to make plaster board?

2　These solid salts were dissolved in 1 dm^3 of distilled water. 25 cm^3 of this solution were titrated against a 1M solution of silver nitrate in the presence of an indicator.

　a　Calculate the volume of silver nitrate solution which would be required for complete reaction with all of the chloride ions present.

　b　Name a suitable indicator used for this titration, as well as the relevant colour change.

3　When the dry sample was heated, the total mass decreased by 0.062 g. A gas A was also produced which gave a white precipitate when bubbled through calcium hydroxide solution.

　a　Identify by name and formula:

　　i　the salt which decomposed on heating,

　　ii　gas A.

　b　Write a balanced chemical equation for the decomposition of the salt you have given as your answer to part **a i**.

　c　Verify by calculation that gas A is in fact the gas you have chosen.

4　Magnesium can be extracted from sea water. The sea water is treated with calcium hydroxide and magnesium hydroxide is precipitated and filtered off. The magnesium hydroxide is then dissolved in hydrochloric acid and the resulting solution is evaporated to produce magnesium chloride. This is then fused, at about 700 °C, with additives which lower its melting point, and electrolysed with a graphite anode and a steel cathode. The magnesium produced is 99.9% pure.

　a　i　Write an ionic equation for the precipitation of magnesium hydroxide.

　　ii　Write a balanced chemical equation for the neutralisation reaction of magnesium hydroxide with hydrochloric acid.

　b　Write anode and cathode electrode equations which take place during the electrolysis of molten magnesium chloride.

36   Water

Pure water, freshly obtained, has a K_W (ionic product for water) of 1×10^{-14} mol^2 dm^{-6} at 25 °C. A student heated a sample of pure water from 20 to 100 °C and determined K_W at various temperatures. The results of these measurements are shown below.

Temperature/°C	K_W/mol^2 dm^{-6}
20	0.68×10^{-14}
25	1.00×10^{-14}
50	5.48×10^{-14}
100	51.30×10^{-14}

1   How does the value of K_W change with increase in temperature ?

2   Deduce the sign of the enthalpy change associated with:

$$H_2O_{(l)} \rightleftharpoons H^+_{(aq)} + OH^-_{(aq)}$$

Explain the reasons for your choice of sign.

3   Calculate the hydrogen ion concentration at 50 °C and hence obtain a value for the pH of the sample of water at this temperature.

4   A further sample of pure water was left to stand in the atmosphere for 30 minutes. It was found to have a pH of 6.3. Account for this pH value.

5   a   A sample of snow was obtained from the French Alps. When this sample was melted and its pH measured it was found to be 4.0. Account for this relatively low pH value.

   b   100 cm^3 of this melted snow were titrated against 0.001M sodium hydroxide.

      i   Calculate the volume of alkali required for complete neutralisation.

      ii   Suggest a suitable indicator which may be used in this titration, and give the colour change.

   c   Write an ionic equation which describes the process which occurs when this snow falls upon buildings constructed of limestone.

37 Chlorine in the Environment

The Mitsubishi Company has developed a new, non–toxic, pollution free, electro–conducting paint which will protect the hulls of ships from encrustation. Sea water contains chloride ions. When a small electric current is passed through the conducting paint on the hull of a ship the chloride ions in contact with the paint are converted to ClO^- ions, which act as a biocide, and prevent encrustation taking place.

1   What is the oxidation state of chlorine in ClO^-?

2   Chlorine, when dissolved in water also produces the $ClO^-_{(aq)}$ ion.
    a   Write a chemical equation for the reaction between chlorine and water.
    b   Suggest a reason why it is not sensible to bubble chlorine around the hulls of ships to prevent encrustation.

3   The ClO^- biocide breaks down very quickly under the action of sunlight and presents no long term health hazard. The products of this decomposition are chloride ions and oxygen.
    a   Write an ionic equation to represent the decomposition of the biocide.
    b   Suggest a method which may be used to follow the rate at which decomposition takes place.

4   $ClO^-_{(aq)}$ is also formed as one of the products of the reaction of chlorine with cold, dilute sodium hydroxide. One of the two other products of this disproportionation reaction is the chloride ion.
    a   Write an equation to represent the reaction outlined above.
    b   What do you understand by the term 'disproportionation' as applied to this reaction?

5   When $NaClO_{(aq)}$ is warmed it decomposes according to the equation:

$$3NaClO_{(aq)} \longrightarrow 2NaCl_{(aq)} + NaClO_{3(aq)}$$

    a   State the oxidation state of chlorine in each of the substances shown in the equation.
    b   State the type of reaction occurring.
    c   Suggest a suitable use for each of the substances shown.

38 Forensic Science

A small amount of a colourless organic liquid containing carbon, hydrogen and oxygen only was found at the scene of a crime which took place in mid–January this year. The mass spectrum obtained for this organic compound is shown below along with its infra–red spectrum. When 1 g of this compound was subjected to combustion analysis, 1.9129 g of carbon dioxide and 1.1736 g of water were produced.

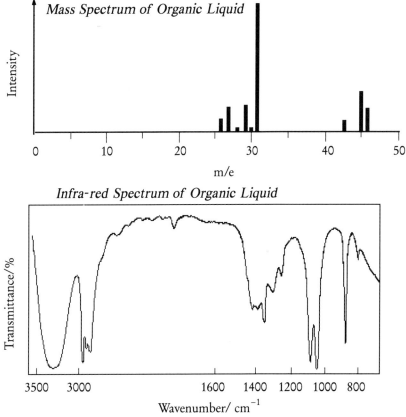

Mass Spectrum of Organic Liquid

m/e

Infra-red Spectrum of Organic Liquid

Wavenumber/ cm⁻¹

1 Use the combustion analysis data to determine the empirical formula of the organic compound. (R.A.M., A_r C=12, H=1, O=16)

2 Use the mass spectrum of the organic compound to obtain its relative molecular mass.

3 a Using your answers to questions **1** and **2**, calculate the molecular formula of the organic compound.

   b Draw two possible structural formulae for this molecular formula.

4 Using the infra–red spectrum of the organic compound and your answers to questions **3 a** and **b**, deduce the actual structure of the substance. Give reasons for your answer.

Some characteristic absorptions of infra–red spectra	
C–OH stretching	$1025-1110$ cm^{-1}
C–H bending	$1400-1450$ cm^{-1}
C–H stretching	$2800-2920$ cm^{-1}
O–H···O stretching	3350 cm^{-1} (broad)

5 Further proof of the identity of an organic compound can be found with reference to a chemical test. Give one chemical test which would help you to confirm your conclusion of the identity of this organic compound. State the reagents and conditions used and the positive result of your test.

39   Lead Pollution

Lead has been added to petrol in small quantities since the early 1920s. At present it is added in the form of tetraethyllead(IV), $(C_2H_5)_4Pb$. The addition of this lead compound prevents pre–ignition and is therefore known as an anti–knock agent. It also acts as an engine lubricant, improving efficiency and economy. Leaded petrol can contain 1.3 g per dm^3 of tetraethyllead(IV).

In order to prevent lead from accumulating in the engine, 1, 2–dibromoethane (CH_2BrCH_2Br) is also added to petrol. This results in the formation of the volatile lead(II) bromide which is expelled in the exhaust gases.

1   Calculate the mass of lead added per dm^3 of petrol.
(R.A.M., A_r Pb=207, C=12, H=1).

2   A car travels 15 000 miles in a year. If the car does on average 9 miles per dm^3 of petrol, calculate the mass of lead released into the environment during the year.

In January 1980, the United Kingdom legislation reduced the permitted lead content to 0.15 g per dm^3 of petrol.

3   Calculate the mass of tetraethyllead(IV) which is now permitted per dm^3 of petrol.

Tetraethyllead(IV) is made by heating chloroethane with a lead–sodium alloy, according to the equation:

$$4CH_3CH_2Cl \ + \ Pb \ + \ 4Na \longrightarrow (C_2H_5)_4Pb \ + \ 4NaCl$$

4   Given that 10 000 tonnes of lead are emitted into the atmosphere from car exhausts per year in the UK, calculate:
    a   the mass of tetraethyllead(IV) added to petrol per year,
    b   the mass of chloroethane required for this manufacture.

5   Chloroethane is manufactured by the reaction of ethene with hydrogen chloride.
    a   State the type of reaction which occurs.
    b   Describe a mechanism for the production of chloroethane by this method.

Corrosion is a set of REDOX processes. The most common example of corrosion is rusting which applies to iron and steel.

Rust is hydrated iron(III) oxide, $Fe_2O_3 . x\, H_2O$. The conditions necessary for rusting to take place are air and water. This process is an electrochemical one involving anode and cathode regions where the following reactions take place:

$$Fe_{(s)} \longrightarrow Fe^{2+}_{(aq)} + 2e^-$$

and

$$O_{2(aq)} + 2H_2O_{(l)} + 4e^- \longrightarrow 4OH^-_{(aq)}$$

1   **a**   What do you understand by the term REDOX process ?

    b   Which of the above is the anode reaction?

    c   Determine an overall REDOX equation for the rusting process.

The next stage of the process is further oxidation of iron(II) hydroxide to hydrated iron(III) oxide. This oxidation is extremely rapid.

2   **a**   Write electronic configurations for iron(II) and iron(III) ions.

    b   Suggest a reason why this oxidation process is 'extremely rapid'.

Rusting costs millions of pounds each year in both replacement of rusted items and the need to protect items from rusting. About 12% of the annual production of steel in the UK is needed simply to replace that lost through rusting! It is not surprising, therefore, that many methods of protection have been developed to inhibit this process. A very successful method (which is known as sacrificial protection) is to coat steel with a more reactive metal.

3   Using the following data,

$$E^{\ominus}/V$$

$$Mg^{2+} + 2e^- \rightleftharpoons Mg \quad -2.37$$

$$Zn^{2+} + 2e^- \rightleftharpoons Zn \quad -0.76$$

$$Fe^{2+} + 2e^- \rightleftharpoons Fe \quad -0.44$$

$$Sn^{2+} + 2e^- \rightleftharpoons Sn \quad -0.14$$

$$Cu^{2+} + 2e^- \rightleftharpoons Cu \quad +0.34$$

decide which metals:

a   would protect iron from rusting,

b   would not protect iron from rusting.

4   Suggest a reason why zinc chromate(VI) is incorporated into primers used to undercoat car bodies.

The problem of corrosion is not confined to the planet earth. Chemical rocket engineers are confronted by major blockage problems associated with dinitrogen tetroxide, N_2O_4, corroding both pipelines and storage tanks.

It is thought that the iron(III) complex:

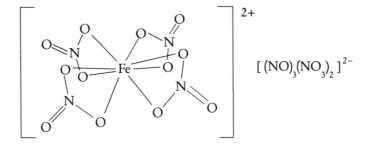

$$[\,(NO)_3(NO_3)_2\,]^{2-}$$

is formed by the reaction of iron with a nitric acid impurity in the dinitrogen tetroxide.

5 a   Name the ligand in this complex ion.

    b   Suggest the type of bonding found within this complex ion.

    c   Write the formula of the iron(II) complex shown.

    d   Suggest a method of reducing the corrosion.

1   CFC 12

1   X = antimony(III) fluoride, SbF_3.

Z = hydrogen chloride, HCl.

2 $3CCl_{4(l)} + 2SbF_{3(s)} \longrightarrow 3CCl_2F_{2(g)} + 2SbCl_{3(s)}$

$6HF_{(g)} + 2SbCl_{3(s)} \longrightarrow 2SbF_{3(s)} + 6HCl_{(g)}$

The first stage usually takes place in the presence of SbF_5.

3   Greenhouse gases – the trapping by CFC 12 molecules of long wave radiations emitted by the earth. Other molecules such as water (vapour) and carbon dioxide as well as methane are also involved. This leads to the temperature at the earth's surface being considerably higher than would otherwise be the case.

Ozone depletion – ozone is produced by a photochemical reaction which converts O_2 into O_3. Ozone can absorb UV light and in so doing keeps most of these harmful rays from penetrating the earth's lower atmosphere. Increasing quantities of man–made substances are attacking this layer and reducing the amount of ozone in it – hence 'ozone depletion'. In 1983 the British Antarctic Survey noted that this was particularly happening over Antarctica in the spring.

4   **a**   Refrigerant.

     b   Aerosol propellant.

5   **a**   A very dramatic increase would have been observed.

     b   Post–protocol should see this increase much reduced.

       Note It is expected that the worst effects of CFCs on the ozone layer will not be seen for a few years as it takes several years for them to reach the upper atmosphere. It has been predicted that if the present rate of CFC production and use continues then the ozone layer may see a 50% reduction by the year 2000.

6   A = chlorine free radicals, $\cdot Cl$.

D = chlorine oxide, ClO.

7 $CCl_2F_2 \xrightarrow{\text{UV}} 2\overset{\bullet}{C}l + \overset{\bullet}{C}F_2$
(CFC 12)

$O_3 + \overset{\bullet}{C}l \longrightarrow ClO + O_2$

then $ClO + \overset{\bullet}{O} \longrightarrow \overset{\bullet}{C}l + O_2$

also $ClO + O_3 \longrightarrow ClO_2 + O_2$

8   A and D are involved in a chain reaction in which chlorine free radicals are regenerated. Hence, one chlorine free radical may cause quite a large amount of ozone conversion.

9   Four bond pairs equally repel; tetrahedral symmetry; bond angles of 109.5°.

2   An Organic Problem

1 C_4H_6O

2 C_4H_6O

3   **a**   X contains an aldehyde or ketone functional group,

     b   X contains a ketone functional group,

     c   X contains a C=C.

4

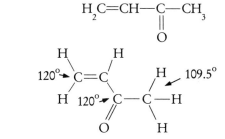

5

6   **a**   Polar bond – in a covalent bond which has been formed between two different atoms with different electronegativities, the bonding electrons are attracted to the more electronegative element. For example C–Cl, C–F bonds.

      These bonds are shown as:

$$\overset{\delta+}{C} - \overset{\delta-}{Cl} \text{ or } \overset{\delta+}{C} - \overset{\delta-}{F}$$

     b   C=O

7   **a** $CH_2CHCOCH_3 + Br_2 \longrightarrow CH_2BrCHBrCOCH_3$

     b

3   Industrial Equilibrium

1   **a**   **i**   At a given temperature increasing the pressure increases the percentage yield of ammonia.

       ii   At a given pressure increasing the temperature decreases the percentage yield of ammonia.

    b   Exothermic reaction = $-\Delta H$

       The answer to part **a ii** indicates that the equilibrium has been displaced to the left at higher temperature, i.e. in a direction to absorb the increased input of heat energy (Le Chatelier's principle), i.e. in the direction of the endothermic reaction. Hence, the production of ammonia will be accompanied by a negative enthalpy change.

2 a 25%

b
$$N_2 + 3H_2 \rightleftharpoons 2NH_3$$

Volume of ammonia produced = 400 000 x 25/100
= 100 000 dm³

Using $PV = nRT$: $n = 336936.87$

Mass = 5727.93kg

c Le Chatelier's principle suggests that the increase in pressure to 400 atmospheres and decrease in temperature to 450 K would increase the percentage of ammonia at equilibrium. However, the use of a higher pressure requires stronger, more expensive equipment with greater maintenance requirements. Also, if the temperature is too low the rate of reaction is so slow that it makes the process uneconomical. The conditions employed are a compromise.

3 a The gases are cooled whilst still under pressure and the ammonia is liquified and removed in this state.

b Removal of the product favours the formation of the product.

4 a i Nitrogen = 20% ; **ii** Hydrogen = 60%

b $K_p = pNH_3{}^2/pN_2.pH_2{}^3$

c pN_2 = 0.2 x 200 = 40 atm.

pH_2 = 0.6 x 200 = 120 atm.

pNH_3 = 0.2 x 200 = 40 atm.

K_p = 2.315 x 10⁻⁵ atm⁻²

d Lower pressure would reduce the partial pressure of ammonia in the equilibrium mixture. The value of K_p would decrease.

e Gases behave in an ideal manner.

5 a The catalyst is used to offset the rate effect observed even with a temperature of 723 K. The best catalyst is iron mixed with small amounts of promoters such as aluminium oxide and potassium oxide (these improve the activity of the catalyst).

b

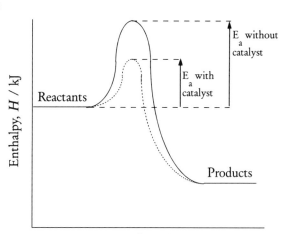

4   Fission/Fusion − 1

1 A = Hydrogen, H_2

D = Uranium(IV) oxide, UO_2

E = Magnesium fluoride, MgF_2

$$UO_{3(s)} + H_{2(g)} \longrightarrow UO_{2(s)} + H_2O_{(g)}$$
$$UO_{2(s)} + 4HF_{(g)} \longrightarrow UF_{4(s)} + 2H_2O_{(g)}$$
$$UF_{4(s)} + 2Mg_{(s)} \longrightarrow U_{(l)} + 2MgF_{2(s)}$$

2 a 1500 moles.

  b 354.43 kg

3 Calcium. This is a reactive metal in plentiful supply. It is not too reactive otherwise very specialised conditions would be required.

4 a 6+

  b 6+

5

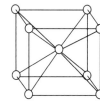

Body-centred cubic structure

6 Uranium conducts electricity in a similar manner to other metals. If a potential difference is applied between the ends of a metal then the delocalised electrons will flow towards the positive potential.

7 a $$U_{(s)} + 2H_2O_{(l)} \longrightarrow UO_{2(s)} + 2H_{2(g)}$$

  b $$2U_{(s)} + 3H_{2(g)} \longrightarrow 2UH_{3(s)}$$

5   Equations of State

1 n = number of moles of gas.

R = molar gas constant.

2 'a/V^2' − Intermolecular forces cannot be neglected. In the bulk of gas they cancel each other out but at the walls of a vessel the molecules about to strike the surface are subject to a force tending to pull them back into the bulk. Therefore, the measured pressure is lower than the theoretical pressure. The size of the pressure reduction is proportional to:

i the concentration of molecules near the wall (which is proportional to $1/V$) and

ii the concentration of molecules within the bulk (which is also proportional to $1/V$).

Hence, the pressure reduction is represented by a term 'a/V^2' where 'a' is a constant called the pressure correction coefficient.

'*V-b*' – Molecules do not have negligible volume. Each one has a finite size which then must exclude a certain volume of a container from all the others. If we call this volume which is excluded *b*, then the actual volume in which the molecules move is *(V-b)*. '*b*' is a constant for a particular gas.

3   At low pressures the molecules are widely spaced, and at high temperatures the molecules are moving rapidly. In both cases the intermolecular forces are not significant. The gas is therefore behaving in the extreme like a gas and least like a liquid.

4   **a** **i** $PV = nRT$

   $P = 15.291125$ atmospheres.

   ii $(P + a/V^2)(V-b) = nRT$

   $P = 15.291132$ atmospheres

   b The small increase in pressure is due to the presence of intermolecular forces.

5   **a**

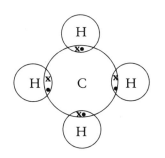

   b Four covalent bond pairs repelling each other equally; tetrahedral symmetry; bond angles of 109.5°.

6   A Friedel–Crafts Catalyst

1   Because the aluminium chloride produced is readily attacked by water it must be prepared under anhydrous conditions.

2   Mass of Al : mass of Cl
      0.54 g : 2.13 g
   Moles of Al : moles of Cl
      0.02 : 0.06
   Hence, $AlCl_3$ ($M_r = 133.5$)

   As solid $M_r = 267$, hence exists as Al_2Cl_6 dimer.

   In benzene solution $M_r = 133.5$, hence exists as $AlCl_3$ monomer units.

3   Hydrogen chloride is formed by the hydrolysis reaction of water with aluminium chloride. The hydrogen chloride dissolves in the water on the indicator paper producing hydrochloric acid.

4   **a** $[Al(H_2O)_6]^{3+}$

b Al^{3+} ion is highly polarising and weakens the OH bond of the coordinated water. Other solvent water molecules can act as bases and abstract protons:

$$[Al(H_2O)_6]^{3+}_{(aq)} + H_2O_{(l)} \rightleftharpoons [Al(H_2O)_5OH]^{2+}_{(aq)} + H_3O^+_{(aq)}$$

$$[Al(H_2O)_5OH]^{2+}_{(aq)} + H_2O_{(l)} \rightleftharpoons [Al(H_2O)_4(OH)_2]^+_{(aq)} + H_3O^+_{(aq)}$$

5   **a and c** In this type of reaction the aluminium chloride (an electron deficient compound) behaves as a Lewis acid. The catalyst withdraws the electrons from the C–Cl bond. The benzene provides a pair of electrons to form a bond to the acyl group.

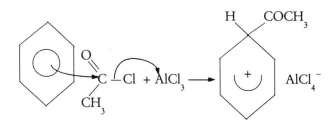

The resulting carbonium ion then reacts with $AlCl_4^-$.

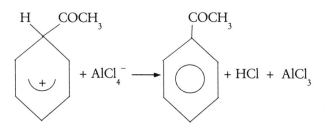

and the $AlCl_3$ catalyst is regenerated.

b $C_6H_6 + CH_3COCl \xrightarrow[\text{catalyst}]{AlCl_3 \text{ as}} C_6H_5COCH_3 + HCl$
      phenylethanone

7   An Essential Metal

1   Solid

2   Calcium

   a Ca

   b 40

   c 20

   d Ca $1s^2\ 2s^2\ 2p^6\ 3s^2\ 3p^6\ 4s^2$
      $Ca^{2+}\ 1s^2\ 2s^2\ 2p^6\ 3s^2\ 3p^6$

   e 2

   f 4

3 a $Ca_{(s)} + 2H_2O_{(l)} \longrightarrow Ca(OH)_{2(aq)} + H_{2(g)}$

                                A   D

           calcium hydroxide   hydrogen

b Ca^{2+} OH^-

4 a $2Ca_{(s)} + O_{2(g)} \longrightarrow 2CaO_{(s)}$

                          E

                 calcium oxide

b $CaO_{(s)} + H_2O_{(l)} \longrightarrow Ca(OH)_{2(aq)}$

5 a $Ca_{(s)} + Cl_{2(g)} \longrightarrow CaCl_{2(s)}$

                            G

             calcium chloride

b

$[Ca]^{2+}$ $\begin{bmatrix} & xx & \\ \bullet & Cl & xx \\ & xx & \end{bmatrix}^-$ $\begin{bmatrix} & xx & \\ \bullet & Cl & xx \\ & xx & \end{bmatrix}^-$

(2, 8, 8)   (2, 8, 8)   (2, 8, 8)

6 a

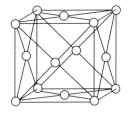

Face-centred cubic structure

b Bonding in metals – the metal atoms in the crystal pool their outer shell electrons (valence electrons) in a common energy level, i. e. the electrons are delocalised. The removal of these electrons leaves metal cations behind. The high electrical conductivity of metals is due to the presence of these 'free' electrons which can move through the metal structure when it is connected to an electrical supply. The electrons flow towards the positive potential.

8   A Genetic Element

1 Solid
2 Phosphorus
   a P
   b 31
   c 15
   d $[Ne]\,3s^2\,3p^3$
   e $(P^{3-})\,[Ne]\,3s^2\,3p^6$
   f 5
   g 3

3 $P_{4(s)} + 5O_{2(g)} \longrightarrow 2P_2O_{5(s)}$

4 a $P_{4(s)} + 6Cl_{2(g)} \longrightarrow 4PCl_{3(l)}$

                                   A

         $PCl_{3(l)} + Cl_{2(g)} \longrightarrow PCl_{5(s)}$

                                   D

   b A : Three covalent bond pairs + one lone pair; pyramidal; bond angle 107°.

        D : Five covalent bond pairs; trigonal bipyramid; bond angles 120° , 90°.

5 PH_3 (phosphine)

    $P_{4(s)} + 3NaOH_{(aq)} + 3H_2O_{(l)} \longrightarrow PH_{3(g)} + 3NaH_2PO_{2(aq)}$

                                                 E

6 a Nitrogen, oxygen, carbon.

   b Nitrogen: (+5) Nitric acid, HNO_3
       Oxygen: (+2) Oxygen difluoride, OF_2
       Carbon: (+4) Carbon dioxide, CO_2

9   Analysis of a Mineral Water

1 Initially concentrate the salts present in the mineral water by boiling.

   a To test for a sulphate – a few drops of $BaCl_{2(aq)}$ and $HCl_{(aq)}$ are added to a sample of the mineral water. If a sulphate is present a white precipitate of $BaSO_{4(s)}$ is formed.

   b The brown ring test will show the presence of a nitrate. To a small amount of the mineral water add some freshly prepared $FeSO_{4(aq)}$ and a few drops of concentrated H_2SO_4. The formation of a ring of the brown complex, $FeSO_4.NO$, will be seen if a nitrate is present.

   c To test for a chloride – a few drops of silver nitrate solution are added to a sample of the mineral water (acidified with a little dilute nitric acid). If a chloride is present, a white precipitate of $AgCl_{(s)}$ is formed. If left to stand the precipitate goes grey.

   Problems would be associated with i) the concentration of the solution and ii) silver sulphate may precipitate during the test for the chloride.

2   **a** +6   **b** +5   **c** −1   **d** +4

3   **a** Trigonal planar: 120°   **b** Tetrahedral: 109.5°

4 Titration using silver nitrate solution – 100 cm³ of mineral water are pipetted into a conical flask. A few drops of potassium chromate(VI) are added. 0.05M silver nitrate solution is titrated until a reddish tinge of silver chromate(VI) appears. It is necessary to carry out a rough titration first due to the difficulty of detecting the end– point. The titration should be repeated until two consistent results are obtained.

$$Ag^+_{(aq)} + Cl^-_{(aq)} \longrightarrow AgCl_{(s)}$$

The moles of chloride ion present are obtained from the amount of siver nitrate solution used. This value can then be converted to a mass and hence to mg per dm^3.

5 a A possibility:

$$HCO_3^-{}_{(aq)} \rightleftharpoons H^+_{(aq)} + CO_3^{2-}{}_{(aq)} \quad \textbf{\textit{(1)}}$$

$$H_2O_{(l)} \rightleftharpoons H^+_{(aq)} + OH^-_{(aq)} \quad \textbf{\textit{(2)}}$$

The CO_3^{2-} present from equilibrium *(1)* becomes involved with the H^+ from equilibrium *(2)*

$$CO_3^{2-}{}_{(aq)} + 2H^+_{(aq)} \rightleftharpoons H_2CO_{3(aq)}$$

This reduces the $[H^+]$ within the sample.

b $pH = -\log_{10}[H^+_{(aq)}]$

$[H^+_{(aq)}]$ is measured in $mol\ dm^{-3}$.

$7.8 = -\log_{10}[H^+_{(aq)}]$

$[H^+] = 6.31 \times 10^{-8}\ mol\ dm^{-3}$

6 Only the hydrogen carbonates of alkali metals can be obtained as solids when evaporated to dryness. These decompose at temperatures above $70^{o}C$.

$$2NaHCO_{3(s)} \longrightarrow Na_2CO_{3(s)} + H_2O_{(g)} + CO_{2(g)}$$

Water and carbon dioxide are therefore lost on heating to dryness and then to $180^{o}C$.

7 a Na [Ne]$3s^1$ $\quad$ Na^+ [Ne]

K [Ar]$4s^1$ $\quad$ K^+ [Ar]

Mg [Ne]$3s^2$ $\quad$ Mg^{2+} [Ne]

Ca [Ar]$4s^2$ $\quad$ Ca^{2+} [Ar]

b Atomic and ionic radius:

i There is an increase down a group. Down the group extra electrons are added to additional energy levels. There is extra nuclear charge but, because of increased screening, there is a net increase in the atomic radius down the group.

ii There is a decrease in any period from left to right. The extra electrons are being added to the same energy level and the extra nuclear charge attracts these electrons and draws them in closer to the nucleus.

First ionisation energy

i There is a decrease down a group. This is due to the outer electron becoming progresively further away from the nucleus. A further effect is that of shielding caused by the extra filled energy levels.

ii There is an increase in any period from left to right. This is due to the increasing nuclear charge making it more difficult to remove an outer electron.

c Na/K – The second ionisation energy would be much greater than the first. This is due to the second electron being taken from an inner filled energy level.

Mg/Ca – The second ionisation energy would be greater than the first. However, the increase would not be so marked because the second electron was not removed from an inner energy level.

10 Fission/Fusion – 2

1 An isotope of uranium, of A_r 235 measured on the ^{12}C scale. 92 represents the atomic number (Z) (number of protons). This number defines the element.

2 It is the induced (or spontaneous) disintegration of the U–235 nucleus into two lighter atoms. This involves a loss of mass which is converted into nuclear energy.

3 $\quad$ $^{92}_{a}X = {}^{92}_{36}Kr$; $\quad$ $^{b}_{56}E = {}^{141}_{56}Ba$

4 A daughter element is a nuclide that originates from the radioactive disintegrations of a parent nuclide.

5 $\quad$ $^{239}_{92}U \longrightarrow {}^{239}_{93}Np + {}^{0}_{-1}e$

$\quad$ $^{239}_{93}Np \longrightarrow {}^{239}_{94}Pu + {}^{0}_{-1}e$

6 The 'fast–breeder' reactor uses a mixture of U–238 and plutonium. The plutonium is sufficiently concentrated for the fission reaction to be sustained and no moderator is used since fast neutrons are needed. The uranium present captures neutrons and changes to plutonium. As the original plutonium is used up, more is created from the U–238. This 'breeder' reactor breeds its own fuel.

7 The energy from the reactor core is removed by carbon dioxide or water at high pressure (Magnox Advanced Gas Cooled Reactor). The hot water or gas boils water, and the steam so produced is used to drive a generator.

8 Fusion is the process of forming new atomic nuclei by the fusion of lighter ones. In this case helium is formed from the isotope of hydrogen (called deuterium).

9 Deuterium.

10 $^{3}_{2}He$.

11 Biochemistry

1 Step a: chlorine gas, UV light. With excess ethanoic acid, chloroethanoic acid predominates.

$$CH_3COOH + Cl_2 \longrightarrow ClCH_2COOH + HCl$$

(Problem: some dichloro– and trichloro– derivatives are obtained.)

Step b: concentrated ammonia followed by addition of dilute hydrochloric acid.

$$ClCH_2COOH + 2NH_3 \longrightarrow H_2NCH_2COO^-NH_4{}^+$$

$$H_2NCH_2COO^-NH_4{}^+ + HCl \longrightarrow H_2NCH_2COOH + NH_4Cl$$

(Problem: The amino group which is introduced can react with a second molecule of chloroacid

$$H_2NCH_2COOH + ClCH_2COOH \longrightarrow HN(CH_2COOH)_2 + HCl\,)$$

2 Step a: initiation:

$$Cl_2 \longrightarrow 2Cl^\bullet$$

propogation:

$$CH_3COOH + Cl^\bullet \longrightarrow {}^\bullet CH_2COOH + HCl$$

$${}^\bullet CH_2COOH + Cl_2 \longrightarrow ClCH_2COOH + Cl^\bullet$$

termination:

e.g. $${}^\bullet CH_2COOH + Cl^\bullet \longrightarrow ClCH_2COOH$$

Step b: nucleophilic substitution

$$H_3N: \underset{\delta-}{Cl}-\underset{\delta+}{CH_2}-COOH \longrightarrow H_3N^+-CH_2-COOH \quad Cl^-$$

$$\longrightarrow NH_2CH_2COOH + HCl$$

3 Peptide –CONH–

4 It is suggested that polypeptides are made of up to 40 alpha amino acid residues. However the distinction between polypeptides is an arbitrary one.

5 Proteins are hydrolysed in the first instance to smaller units such as dipeptides and tripeptides, and then to alpha amino acids under acid conditions by pepsin, an enzyme found in the stomach, as well as (at pH 6–8) chymotrypsin and trypsin, enzymes found in the intestines. The amino acids so produced then pass into the blood stream where they can travel to organs such as the liver and other tissues where under the influence of nucleic acids they are converted into proteins which are required by our bodies.

6 Nylon–6.6 (a polyamide)

7 $$HOOC-CH_2CH_2CH_2CH_2-COOH$$
(hexanedioic acid)

8 Dilute hydrochloric acid.

$$H_2NCH_2CH_2CH_2CH_2CH_2CH_2NH_2$$
(1,6–diaminohexane)

9 a This polymer carries highly polar groups along its chain:

$$\begin{array}{cc} O & H \\ \| & | \\ -C- & N- \end{array}$$

b H–bonding takes place between adjacent chains. These relatively strong intermolecular interactions tend to increase the crystallinity of this polymer.

12 Graphitic Compounds

1 a Allotropes of carbon – Elements in two or more solid, liquid or gaseous forms, in one phase are called allotropes. Graphite is one of the two allotropes of carbon, the other being diamond.

b Covalent bonds – These are chemical bonds in which two or more atoms (usually non–metals) are held together by the overlap of atomic orbitals. There is a strong interaction of the nucleii of these atoms with one or more pairs of electrons. In graphite, each carbon atom is joined by a single covalent bond to three other atoms in its layers. The bonding in graphite can be visualised as three trigonally–arranged covalent bonds. These covalent bonds are formed by three of the four valence electrons of carbon.

c Unsaturation – The carbon–carbon bond length within a layer is 0.142 nm. This bond length is intermediate in length between a single and double carbon–carbon bond which suggests that there is some multiple bond character present.

d Non–stoichiometric – Inexact proportions of elements present in the compound.

e Macromolecular – This is a term used in connection with very large molecules containing greater than 10 000 atoms. The covalent bonding in boron nitride and graphite extends from one atom to the next through the whole lattice, forming a three–dimensional giant molecule.

2 Van der Waals

3 Cs $[Xe]\,6s^1$

4 In Caesium, the outer shell electron is relatively easily removed to form the caesium cation. These outermost electrons then move freely within the metal lattice. Therefore, the metal consists of a three–dimensional array of positive ions surrounded by a 'sea' of electrons. These outer electrons are delocalised. The structure is held together because each cation is attracted to the delocalised electron cloud between it and the next cation.

5 The fourth electron in the outer shell of carbon is delocalised throughout the layer structure. This delocalisation results in graphite conducting electricity. The introduction of further delocalised electrons from added caesium enhances the electrical conductivity of the substance created.

6 It does not possess delocalised electrons.

7 Neutrons which result from the fission of U−235 are slowed down by collision with the carbon atoms of graphite. By this method it ensures that as many neutrons as possible collide and cause further fission of U−235 atoms. It should be noted that neutrons are not absorbed by collision with the carbon atoms of graphite. If the temperature of the reactor increases to too high a level then movable absorbing rods of cadmium or boron may be lowered. By carefully adjusting the position of these rods the chain reaction can be controlled.

8 High pressure and temperature would displace the equilibrium to the right in favour of C (diamond).

13   Effusion

1
$$\frac{R(X)}{R(C_4H_{10}O)} = \sqrt{\frac{M(C_4H_{10}O)}{M(X)}}$$

$$M(X) = 88.5$$

2 Possible empirical formula: $C_4H_8O_2$
Isomers: $CH_3CH_2COOCH_3$
$CH_3COOCH_2CH_3$
$CH_3CH_2CH_2COOH$

3 This ester contains the grouping $-COCH_3$

$$X = CH_3CH_2COOCH_3$$

4 a Graham's Law

b
$$\frac{R(^{235}UF_6)}{R(^{238}UF_6)} = \frac{M(^{238}UF_6)}{M(^{235}UF_6)}$$

Therefore, ratio is 1.006.

14 Relative Molecular Mass

1
$$PV = (m/M).RT$$

$$M = \frac{mRT}{PV}$$

$$M = 57.99 \text{ g mol}^{-1}$$

2
$$X = CH_3COCH_3$$
Isomeric aldehyde $= CH_3CH_2CHO$

3 a $Y = CH_3CH(OH)CH_3$ Reduction

b $CH_3CH_2CH_2OH$

4
$$CH_3 - \underset{\underset{\displaystyle C_6H_5(NO_2)_2}{\overset{\displaystyle |}{\underset{\displaystyle |}{\overset{\displaystyle N}{\overset{\displaystyle \|}{C}}}}}{\overset{\displaystyle N H}{}} - CH_3$$ Condensation

5 Addition of ammoniacal silver nitrate (Tollens' reagent). On warming, a silver mirror is observed in the case of the aldehyde. Ag^+ is reduced to Ag.

15 Anti−Cancer Drug

1 Carboxyl group, COOH
Halogenoalkane, C−Cl

2 Extends the useful action (and persistence) of the drug.

3 a
$$Na^{+-}O-\overset{\overset{\displaystyle O}{\|}}{C}-(CH_2)_3-C_6H_5-N(CH_2CH_2OH)_2$$

b
$$C_2H_5O-\overset{\overset{\displaystyle O}{\|}}{C}-(CH_2)_3-C_6H_5-N(CH_2CH_2Cl)_2$$

c
$$HO-\overset{\overset{\displaystyle O}{\|}}{C}-(CH_2)_3-C_6H_5-N(CH_2CH_2CN)_2$$

4 a
$$---\overset{\overset{\displaystyle H}{|}}{\underset{\underset{\displaystyle H\quad CN^-}{}}{C^+}}{\overset{\delta+}{\cdots}}{\overset{\delta-}{}}Cl^- \longrightarrow ---\overset{\overset{\displaystyle H}{|}}{\underset{\underset{\displaystyle H}{|}}{C}}-CN + Cl^-$$

b Nucleophilic substitution.

5 a
$$pK_a = -\log_{10} K_a$$
$$4.9 = -\log_{10} K_a$$
$$K_a = 1.26 \times 10^{-5} \text{ mol dm}^{-3}$$

b Chlorambucil behaves as an aliphatic acid. Benzoic acid is a stronger acid because the negative charge on the carboxyl group is delocalised with the pi electrons of the benzene ring.

16   Iron

1   X = Carbon dioxide, CO_2

    Z = Carbon monoxide, CO

$$C_{(s)} + O_{2(g)} \longrightarrow CO_{2(g)}$$

$$CO_{2(g)} + C_{(s)} \longrightarrow 2CO_{(g)}$$

$$Fe_2O_{3(s)} + 3CO_{(g)} \longrightarrow 2Fe_{(l)} + 3CO_{2(g)}$$

2 ΔH_1 is exothermic (-392 kJ mol^{-1})

    (Note – ΔH_2 is also slightly exothermic.)

3   A = Calcium oxide, CaO

$$CaCO_{3(s)} \longrightarrow CaO_{(s)} + CO_{2(g)}$$

$$CaO_{(s)} + SiO_{2(s)} \longrightarrow CaSiO_{3(l)}$$

4 $[Fe(H_2O)_6]^{2+}$.

5   A pale green solution would give a pale green gelatinous precipitate which would go brown on standing. A yellow solution would be produced at the surface.

6   The pale green gelatinous precipitate is iron(II) hydroxide. The green colour is due to the presence of iron(III) hydroxide which has been produced either by some iron(III) ions present in the iron(II) solution or by aerial oxidation. Standing in air converts iron(II) hydroxide by oxidation to iron(III) hydroxide (brown).

    (Note – pure iron(II) hydroxide is white.)

7 $5Fe^{2+}_{(aq)} + MnO_4^-{}_{(aq)} + 8H^+_{(aq)} \longrightarrow 5Fe^{3+}_{(aq)} + Mn^{2+}_{(aq)} + 4H_2O_{(l)}$

8   Moles $MnO_4^- = 5 \times$ moles $Fe^{3+} = 5(25/1000 \times 0.02)$

       Mass Fe $= 56 \times 5(25/1000 \times 0.02) = 0.14$ g

       %Fe $= 0.14/0.2 \times 100 = 70\%$

9   In a complex ion a ligand is the ions, molecules or atoms which surround the central atom, e.g. H_2O in $[Fe(H_2O)_6]^{2+}$ or CN^- in $[Fe(CN)_6]^{4-}$. The haem unit is a tetradentate ligand which means it has as a ligand, four donor points to the central Fe^{2+} ion. This coordination takes place from the nitrogen atoms within the ligand. Incidentally it is also from two nitrogen atoms in the globin.

10 a/b One oxygen molecule can be weakly and reversibly attached to one of the globin nitrogen sites. (Note – much stronger ligands such as cyanide or CO can replace the oxygen molecules.)

11   Coordination takes place through the nitrogen atoms. It is by a dative (or coordinate) covalent bond.

12   Anaemia – this is the reduction of the total amount of circulating haemoglobin in the blood.

17   An Analgesic

1   Phenolic group, OH
    Amide group, NH–CO–R

2   There will also be a reaction between the –OH group and ethanoyl chloride to produce an ester.

    $H_2NC_6H_5OH + CH_3COCl \rightarrow CH_3COOC_6H_5NH_2 + HCl$

3   **a**   If excess analgesic:

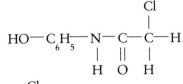

       If excess Cl_2:

    b

    c

4   **a**   If excess analgesic is present:

       initiation:

$$Cl_2 \longrightarrow 2Cl^\bullet$$

       propogation:

$$--CH_3 + Cl^\bullet \longrightarrow --{}^\bullet CH_2 + HCl$$

$$--{}^\bullet CH_2 + Cl_2 \longrightarrow --CH_2Cl + Cl^\bullet$$

       termination: e.g.

$$--{}^\bullet CH_2 + Cl^\bullet \longrightarrow --CH_2Cl \ etc$$

    b   Free radical substitution.

5   Iodoform reaction – iodine in the presence of sodium hydroxide solution. Production of a yellow precipitate is expected if CH_3CO- is present

6 **a** $pK_a = -\log_{10} K_a$

  b $9.5 = -\log_{10} K_a$

  $K_a = 3.162 \times 10^{-10} \text{ mol dm}^{-3}$

  c Due to the presence of the phenolic $-OH$. This creates the presence of phenoxide type ion with the negative charge delocalised around the ring.

18 Titanium

1 R = Chlorine, Cl_2

  X = Titanium(IV) chloride, $TiCl_4$

  Z = Magnesium chloride, $MgCl_2$

$$TiO_{2(s)} + C_{(s)} + 2Cl_{2(g)} \longrightarrow TiCl_{4(l)} + CO_{2(g)}$$

$$TiCl_{4(l)} + 2Mg_{(s)} \longrightarrow Ti_{(s)} + 2MgCl_{2(s)}$$

2 Sodium.

3 By fractional distillation.

4 Traces of air make titanium metal brittle.

5 Supersonic aircraft and space vehicle construction.

6

$$Ti_{(s)} + O_{2(g)} \longrightarrow TiO_{2(s)}$$

$$Ti_{(s)} + 2H_2O_{(g)} \longrightarrow TiO_{2(s)} + 2H_{2(g)}$$

7 **a** Titanium is a transition element because it can form compounds in which there is an incomplete subshell of $3d$ electrons. It also shows variable oxidation states and produces coloured compounds. Zinc is not normally regarded as a transition element since it has a full $3d$ subshell as an element as well as in compound formation. Also, zinc does not show variable oxidation states or produce coloured compounds.

  b **i** Ti $[Ar]\, 3d^2\, 4s^2$

  ii Ti^{2+} $[Ar]\, 3d^2$

  iii Ti^{4+} $[Ar]$

8 **a** A = Zinc carbonate, $ZnCO_3$

  D = Carbon dioxide, CO_2

  E = Zinc oxide, ZnO

  b $[Zn(OH)_4]^{2-}$

  c $ZnO_{(s)} + H_2SO_{4(aq)} \longrightarrow ZnSO_{4(aq)} + H_2O_{(l)}$

  d The majority of the complexes of zinc have a coordination number of 4.

  $[Zn(H_2O)_4]^{2+}$

  e Amphoteric means having both basic and acidic properties. Zinc oxide forms salts with both sodium hydroxide and dilute sulphuric acid.

19 Reactions of Anaesthetics

1 Amine group, NH_2
Ester group, OR

2 **a** $Cl^- N_2^+ -C_6H_5-\overset{\displaystyle O}{\overset{\displaystyle \|}{C}}- O -CH_2-CH_2- N(CH_2CH_3)_2$

  b Diazotisation. Other organic products associated with this reaction which may take place between the hydrochloric acid and the ester linkage producing:

  $$Cl^- N_2^+ - C_6H_5 - \overset{\displaystyle O}{\overset{\displaystyle \|}{C}} - O - H$$

  and $HO-CH_2CH_2N(CH_2CH_3)_2$

  c

  $$HO-C_6H_5-N=N -C_6H_5-\overset{\displaystyle O}{\overset{\displaystyle \|}{C}}-O- (CH_2)_2 - N (CH_2CH_3)_2$$

  Other organic products associated with this reaction are formed by the alkaline conditions affecting the ester linkage producing:

  $$HO - C_6H_5 - N = N - C_6H_5 - \overset{\displaystyle O}{\overset{\displaystyle \|}{C}} - O^- Na^+$$

  and $HO - CH_2CH_2N(CH_2CH_3)_2$

3 **a** $2N$ atoms are present. Both potentially have lone pairs of electrons to donate.

  b $pK_b = -\log_{10} K_b = -\log_{10} 4.1 \times 10^{-7} = 6.39 \text{ mol dm}^{-3}$

4 1−bromo−1−chloro−2−trifluoroethane.

5 Nucleophilic substitution is the displacement of the halogen from the electropositive carbon centre at the other end of the carbon−halogen polar bond, such as C−Cl, by a nucleophile such as OH^-.

6 **a** C−F

  b C−I ; C−Br ; C−Cl ; C−F
  easiest   hardest

  c

  $$\begin{array}{ccc} & F & OH \\ & | & | \\ F\!-\!C\!-\!&\!C\!-\!H \\ & | & | \\ & F & OH \end{array}$$

  The C−F bond is so strong that they are extremely unreactive.

20 Tackling Sleeping Sickness

1 a Sodium nitrate(III) + concentrated hydrochloric acid at 5 °C.

b The solution from part **a** at 5 °C added dropwise with stirring to DFMO at 5 °C.

c Bubbles of nitrogen gas.

d

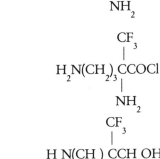

$$H_2NCH_2CH_2CH_2\overset{\overset{\displaystyle CF_3}{|}}{\underset{\underset{\displaystyle NH_2}{|}}{C}}COOH + 2HONO$$

$$\searrow$$

$$HO(CH_2)_3\overset{\overset{\displaystyle CF_3}{|}}{\underset{\underset{\displaystyle OH}{|}}{C}}COOH + 2N_2 + 2H_2O$$

e Follow the rate of evolution of nitrogen with time.

2 It possesses a chiral carbon atom (adjacent to COOH group).

3 Carboxyl group, COOH
Halogenoalkane, C–F

4 a

$$H_2N(CH_2)_3\overset{\overset{\displaystyle CF_3}{|}}{\underset{\underset{\displaystyle NH_2}{|}}{C}}COOC_2H_5$$

b

$$H_2N(CH_2)_3\overset{\overset{\displaystyle CF_3}{|}}{\underset{\underset{\displaystyle NH_2}{|}}{C}}COCl$$

c

$$H_2N(CH_2)_3\overset{\overset{\displaystyle CF_3}{|}}{\underset{\underset{\displaystyle NH_2}{|}}{C}}CH_2OH$$

5 M_r(DFMO) = 182;

M_r(Na salt of DFMO) = 188

$$\text{---COOH} + NaOH \longrightarrow \text{---COONa} + H_2O$$
182 g   188 g
1.82 g   1.88 g

% purity = 1.82/1.88 x 100 = 96.8%

21 Enthalpy of Combustion

1 ΔH = mass x specific heat capacity x rise in temperature

= 500 x 4.2 x 22.9

= 48 090 J (48.09 kJ)

2 Number of moles of butan–1–ol burned

= (29.42–27.51)/74

= 0.0258 moles

Heat given out by one mole = 48.09/0.0258
= 1863.95 kJ

$$\Delta H_c = -\ 1863.95 \text{ kJ mol}^{-1}$$

3 a 30.2%

Possible sources of error are: heat losses due to air currents, neglecting the specific heat capacity of the copper calorimeter, heat escaping from the copper calorimeter (no lid).

b See graph – page 64.

4 –3300 kJ mol^{-1}. Constant increase with the addition of a further –CH$_2$– unit per alcohol molecule.

5 Energy released increases per mole of alcohol burned. The enthalpy change becomes more exothermic.

22 Reaction Kinetics

1 and 2 (All concentrations in mol dm^{-3} x 10^{-2})

Experiment	[I$^-$]	[H$^+$]	[H$_2$O$_2$]	Rate (1/t s^{-1}) x 10^{-2}
1	2.1	2.1	2.1	1.78
2	1.4	2.1	2.1	1.18
3	0.7	2.1	2.1	0.57
4	2.1	1.4	2.1	1.78
5	2.1	0.7	2.1	1.78
6	2.1	2.1	1.4	1.15
7	2.1	2.1	0.7	0.56

3 See graphs – page 64.

4 The reaction rate depends on [I$^-$] and [H$_2$O$_2$].

5 a 1   **b** 0   **c** 1
Rate = k[I$^-$][H$_2$O$_2$]

6 2nd order.

7

$$H_2O_2 + I^- \xrightarrow{\text{slow}} HOI + OH^-$$

$$HOI + I^- \xrightarrow{\text{fast}} I_2 + OH^-$$

$$2OH^- + 2H^+ \xrightarrow{\text{fast}} 2H_2O$$

Overall reaction:

$$H_2O_2 + 2H^+ + 2I^- \longrightarrow I_2 + 2H_2O$$

23 Newly Discovered Elements

1 Time taken for half the nuclei of a sample of a radioactive nuclide to decay.

2 $T_{1/2}$ = 34 s (see graph page 65).

3 5×10^{-4} g

4 a i alpha particle M_r = 4 ; relative charge +2.
   ii beta particle M_r = 0 ; relative charge of −1.

b $^{263}_{106}Unh \longrightarrow ^{259}_{104}X + ^{4}_{2}alpha \qquad X = ^{259}_{104}Unq$

$^{259}_{104}Unq \longrightarrow ^{259}_{105}Y + ^{0}_{-1}beta \qquad Y = ^{259}_{105}Unp$

$^{259}_{105}Unp \longrightarrow ^{255}_{103}Z + ^{4}_{2}alpha \qquad Z = ^{255}_{103}Lw$

5 $^{263}_{106}Unh$: 157 neutrons.

$^{264}_{105}Unp$: 159 neutrons.

6 The RAM is a statistical average based on the percentage abundance and mass of each of the isotopes of the element which are present.

24 Sulphuric Acid

1 A = Sulphur dioxide (sulphur(IV) oxide), SO_2

B = Vanadium(V) oxide, V_2O_5

C = Sulphur trioxide (sulphur(VI) oxide), SO_3

D = Water, H_2O

E = Iron(II) sulphate, $FeSO_4$

F = Hydrogen, H_2

G = Ethene, C_2H_4

H = Poly(ethene) (polythene), $\{C_2H_4\}_n$

I = Hydrogen chloride, HCl

J = Sulphurous acid, H_2SO_3.

2 a $\qquad 2SO_{2(g)} + O_{2(g)} \rightleftharpoons 2SO_{3(g)}$

   b i Pressure – High pressures favour the forward reaction. This is due to the reduction in the number of moles of gas as the reaction proceeds. The overall effect is a reduction in pressure (Le Chatelier's Principle). 1–2 atmospheres are used since a high yield is obtained without the need for higher pressures and, hence, a more expensive plant.

      ii Temperature – Low temperatures would favour the forward, exothermic reaction. However, at low temperatures the reaction rate is slow and so an optimum temperature of 450–500 ° C is used.

iii Catalyst – A catalyst of vanadium(V) oxide is used to decrease the time required for the equilibrium to be attained, by providing an alternative reaction route, with lower activation energy.

3 Dehydrating agent.

4 a Addition polymerisation.

   b Free radical chain reaction.

5 $[Fe(H_2O)_6]^{2+}_{(aq)}$

6 Add bromine water to a container of gas G (ethene). If G (ethene) is present there will be a colour change of orange/red to a colourless solution. This is due to an electrophilic addition reaction occurring between bromine and G (ethene).

7 a With Br^-:

$Br^- + H_2SO_4 \longrightarrow HBr + HSO_4^-$

$2HBr + H_2SO_4 \longrightarrow Br_2 + 2H_2O + SO_2$

With I^-:

$I^- + H_2SO_4 \longrightarrow HI + HSO_4^-$

$2HI + H_2SO_4 \longrightarrow I_2 + 2H_2O + SO_2$

$8HI + H_2SO_4 \longrightarrow H_2S + 4H_2O + 4I_2$

A red gas (Br_2) is produced with a bromide. A violet vapour (I_2) is produced with an iodide.

   b It acts as a non–volatile acid initially and then as an oxidising agent.

8 a $SO_{2(g)} + H_2O_{(l)} \longrightarrow H_2SO_{3(aq)}$
   $H_2SO_{3(aq)} + [O] \longrightarrow H_2SO_{4(aq)}$
   from the air

   b Desulphurisation of flue gases from power stations.

25 Ammonia

1 $2NH_4Cl + Ca(OH)_2 \longrightarrow CaCl_2 + 2NH_3 + 2H_2O$

2 Calcium oxide.

3 Mass of nitric acid required = 39.375 kg.

4 a A Brønsted–Lowry base is a proton acceptor.
   b A Lewis base is an electron pair donor.

5
Approx	1	2	3
20.50	20.20	20.20	20.30

6 Average titre = 20.23 cm^3.

7 a $NaOH_{(aq)} + HNO_{3(aq)} \longrightarrow NaNO_{3(aq)} + H_2O_{(l)}$

   b One mole reacts with one mole.

8 a Number of moles of NaOH in 25 cm^3 = 7.5×10^{-3}

   b 7.5×10^{-3} moles of nitric acid were neutralised.

9 Molarity of nitric acid = 0.37M.

10 Explosives, dyes, polymers.

26 Chloride Analysis

1 The end-point is shown by a change from a white precipitate to a pink precipitate of silver chromate (Ag_2CrO_4).

2 $\quad Ag^+_{(aq)} + Cl^-_{(aq)} \longrightarrow AgCl_{(s)}$

3 1 mole.

4 Number of moles of $Ag^+_{(aq)} = 30 \times 0.05/1000$
$\qquad\qquad\qquad\qquad = 1.5 \times 10^{-3}$

5 Number of moles of $Cl^-_{(aq)}$ = number of moles of $Ag^+_{(aq)}$
Therefore, number of moles of $Cl^-_{(aq)} = 1.5 \times 10^{-3}$

6 Mass of $Cl^-_{(aq)}$ in 25 cm^3 = $1.5 \times 10^{-3} \times 35.5$
$\qquad\qquad\qquad\qquad = 0.0533$ g

7 Mass of $Cl^-_{(aq)}$ in 250 cm^3 = 0.0533×10
$\qquad\qquad\qquad\qquad = 0.533$ g

8 Mass of Group(II) element = $0.85 - 0.533$
$\qquad\qquad\qquad\qquad = 0.317$ g

9 Empirical formula is MCl_2 (M = Group(II) element).

10 Possible sources of error: accurate observation of end-point, accurate reading of volume measurements.

11 a $[Ag(NH_3)_2]^+$

 b $[H_3N \longrightarrow Ag \longleftarrow NH_3]^+$ Bond angle 180°.

12 i Weigh out equal amounts of the two brands of crisps.
 ii Dissolve the salt by adding distilled water (NOT tap water).
 iii Filter.
 iv Pipette 25 cm^3 of the salt solution.
 v Add 3-4 drops of potassium chromate(VI) indicator.
 vi Titrate the salt solution with a suitable strength of silver nitrate solution.
 vii The end-point is when the first pink colouration is observed.
 viii Carry out the calculation as performed earlier.

27 Household Bleaches

1 A = OH^- hydroxide ion,

 B = Cl^- chloride ion.

2 $\quad Cl_{2(g)} + 2OH^-_{(aq)} \longrightarrow Cl^-_{(aq)} + ClO^-_{(aq)} + H_2O_{(l)}$

3 $Cl_{2(aq)} + 2I^-_{(aq)} \longrightarrow 2Cl^-_{(aq)} + I_{2(aq)}$

4 $2S_2O_3{}^{2-}_{(aq)} + I_{2(aq)} \longrightarrow S_4O_6{}^{2-}_{(aq)} + 2I^-_{(aq)}$

5 Number of moles of $S_2O_3{}^{2-}_{(aq)} = 0.1 \times 24.17/1000$
$\qquad\qquad\qquad\qquad\qquad = 2.417 \times 10^{-3}$
Therefore, number of moles of $Cl_2 = 2.417 \times 10^{-3}/2$
$\qquad\qquad\qquad\qquad\qquad = 1.209 \times 10^{-3}$

6 1.209×10^{-2} moles of free chlorine in 250 cm^3 of chlorine solution.

7 Concentration of diluted bleach = $1.209 \times 10^{-2} \times 4 \times 71$
$\qquad\qquad\qquad\qquad\qquad = 3.43$ g dm^{-3}

8 Since the bleach was diluted exactly 25 times, the concentration of the original bleach = 3.43×25 g dm^{-3}
$\qquad\qquad\qquad\qquad\qquad\qquad\qquad = 85.75$ g dm^{-3}

Therefore, the % of available chlorine = 8.575%

9 The HCl would be oxidised giving Cl_2.

10 +1

11 They are environmentally friendly and so do not interfere with natural systems. A feature of hydrogen peroxide as a bleaching agent is that the only by-product of its action is water.

28 Aspirin

1 A = 2-hydroxybenzenecarboxylic acid
B = ethanoic anhydride or ethanoyl chloride.

2 $HOC_6H_4COOH + CH_3COCl$
$\qquad \longrightarrow CH_3COOC_6H_4COOH + HCl$

3 Paracetamol.

4 CH_3CO^- : use the iodoform reaction (triiodomethane), I_2/OH^-. A yellow precipitate of CHI_3 will be observed.
$-COOH$: add sodium carbonate solution. Effervescence would be observed as carbon dioxide is given off. (Damp blue litmus would turn red.)

5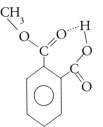

6 Minimal reaction of phenolic $-OH$ assumed

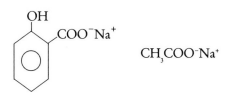

$CH_3COO^-Na^+$

7 In the intestines.

8 $CH_3COOC_6H_4COOH + 2NaOH$

$$\searrow$$

$HOC_6H_4COO^-Na^+ + CH_3COO^-Na^+ + H_2O$

(Note This equation has been produced assuming:
i complete hydrolysis of the ester functional group,
ii little or no reaction with the phenolic $-OH$.

9 Number of moles of NaOH added to the aspirin tablets
$$= 10 \times 1/1000$$
$$= 1 \times 10^{-2}$$

10 $\quad$ $NaOH + HCl \longrightarrow NaCl + H_2O$

11 Number of moles of HCl added to excess NaOH

$$= 25.50 \times 0.1/1000$$
$$= 2.55 \times 10^{-3}$$

Therefore, number of moles of NaOH reacting with the aspirin
$$= 1 \times 10^{-2} - 2.55 \times 10^{-3}$$
$$= 0.00745$$

12 Number of moles of 2−ethanoylhydroxybenzoic acid which have reacted = 0.003725

13 Mass of 2−ethanoylhydroxybenzoic acid
$$= 180 \times 0.003725$$
$$= 0.6705 \text{ g}$$

14 % of 2−ethanoylhydroxybenzoic acid = 0.6705 × 100/1.5
$$= 44.7\%$$

15 Calcium carbonate.

29 $\quad$ A Useful Salt

1 1st ionisation energy of potassium: $K_{(g)} \longrightarrow K^+_{(g)} + e^-$
Enthalpy of formation of $KCl_{(s)}$: $K_{(s)} + 1/2Cl_{2(g)} \longrightarrow KCl_{(s)}$
Atomisation of potassium: $K_{(s)} \longrightarrow K_{(g)}$
1st electron affinity of chlorine: $Cl_{(g)} + e^- \longrightarrow Cl^-_{(g)}$
Atomisation of chlorine: $1/2Cl_{2(g)} \longrightarrow Cl_{(g)}$

2

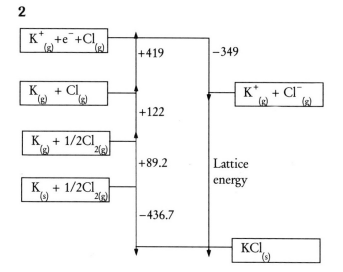

3 Lattice energy for $KCl_{(s)} = -717.9$ kJ mol^{-1}

4 **a** **i** K $[Ne] 3s^2 3p^6 4s^1$ $\quad$ **ii** Cl $[Ne] 3s^2 3p^5$

$\quad$ **iii** K^+ $[Ne] 3s^2 3p^6$ $\quad$ **iv** Cl^- $[Ne] 3s^2 3p^6$

b

$[K]^+$ $\qquad\qquad \begin{bmatrix} & xx & \\ \bullet\, & Cl & xx \\ x & & \\ & xx & \end{bmatrix}^-$

$\quad$ (2, 8, 8) $\qquad\qquad$ (2, 8, 8)

5 **a** $KCl_{(aq)}$ and $AgNO_{3(aq)}$: white precipitate of AgCl.
$KCl_{(s)}$ and concentrated H_2SO_4: white fumes of hydrogen chloride.

$\quad$ **b** $Ag^+_{(aq)} + Cl^-_{(aq)} \longrightarrow AgCl_{(s)}$

$KCl_{(s)} + H_2SO_{4(l)} \longrightarrow KHSO_{4(aq)} + HCl_{(g)}$

6 **a** Double Salt− A salt formed by crystallisation from a solution of two or more components. The resulting salt has distinct properties from the components, as a solid, but has a mixture of the properties of the components as a solution.

Hydrate− A salt which has crystallised in such a form as to retain water molecules within its structure, i.e. water of crystallisation.

$\quad$ **b** Empirical formula: $KCl.MgCl_2.6H_2O$
Mass of 1 mole = 277.5 g.
Molecular formula: $KCl.MgCl_2.6H_2O$

30 A Versatile Transition Metal

1. M=Copper.
2. M^+ (Cu^+) $1s^2\,2s^2 2p^6\,3s^2 3p^6 3d^{10}$
   M^{2+} (Cu^{2+}) $1s^2\,2s^2 2p^6\,3s^2 3p^6 3d^9$
3. CuO
4. a. $[Cu(H_2O)_6]^{2+}$
   b.
   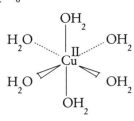

   c. Shape–Octahedral: 6 bonding pairs of electrons surrounding the central metal ion. They repel one another and move as far away from one another as possible to minimise the repulsion. This is achieved when they point towards the corners of an octahedron.

   d. Ligands.

   e. Hexaquocopper(II) ion.

5. a. A white precipitate is formed (CuI) in a brown solution ($I_{2(aq)}$).
   b. $2Cu^{2+}_{(aq)} + 4I^-_{(aq)} \longrightarrow 2CuI_{(s)} + I_{2(aq)}$
6. a. $Cu^+_{(aq)} + e^- \rightleftharpoons Cu_{(s)}$ $E^\ominus = +0.52V$
   $Cu^+_{(aq)} \rightleftharpoons Cu^{2+}_{(aq)} + e^-$ $E^\ominus = -0.15V$
   $2Cu^+_{(aq)} \rightleftharpoons Cu_{(s)} + Cu^{2+}_{(aq)}$ $E^\ominus_{Total} = +0.37V$

   b. Disproportionation, i.e. simultaneous oxidation and reduction of the same species.

   c. The positive $E^\ominus_{Total}$ of the above reaction implies that hydrated M^+ (Cu^+) are thermodynamically unstable in solution with respect to hydrated M^{2+} (Cu^{2+}) and M (Cu).

31   Natural Gas

1. $CO_2 + 2KOH \longrightarrow K_2CO_3 + H_2O$
2. $CO_{2(g)}$ forms an acidic solution of pH 5 when it dissolves in water:
   $$CO_{2(g)} + H_2O_{(l)} \rightleftharpoons H_2CO_{3(aq)}$$
   carbonic acid

The acid conditions would increase the rate at which corrosion takes place.

3. a. $CH_{4(g)} + 2O_{2(g)} \longrightarrow CO_{2(g)} + 2H_2O_{(l)}$
   b. CO_2 does not support combustion.

4. Because $CO_{2(g)}$ becomes a solid at $-33°C$ and would cause blockages of the pipes.

5. $CO_{2(g)}$, an acidic gas, would react with ammonia, an alkaline gas.

6. $H_2S_{(g)} + ZnO_{(s)} \longrightarrow ZnS_{(s)} + H_2O_{(l)}$

7. $CH_{4(g)} + H_2O_{(l)} \overset{\text{Ni catalyst}}{\rightleftharpoons} CO_{(g)} + 3H_{2(g)}$

8. A reducing agent, removing the oxygen from water.

9. a. Increase the concentration of $CH_{4(g)}$ or $H_2O_{(g)}$.
   b. Remove the $CO_{(g)}$ from the equilibrium mixture. The equilibrium moves in a direction to offset the change made.

32 Internal Combustion Engine

1. $\Delta H^\ominus_{r\ reaction} = \sum \Delta H^\ominus_{f\ products} - \sum \Delta H^\ominus_{f\ reactants}$
   $= [(16 \times -394) + (18 \times 242)] - [(2 \times -250) + (0)]$
   $= -10\,160\ kJmol^{-1}$

2. a. $2C_8H_{18(g)} + 17O_{2(g)} \longrightarrow 16CO_{(g)} + 18H_2O_{(g)}$
   b. $\Delta H^\ominus_r = [(16 \times -111) + (18 \times -242)] - [(2 \times -250) + (0)]$
   $= -5632\ kJmol^{-1}$

   c. Different numbers and types of bonds are being broken and formed.

   d. When inhaled, CO combines with haemoglobin in blood to form carboxyhaemoglobin which is a very stable species. This reduces the amount of haemoglobin which is free to transport oxygen, as oxyhaemoglobin, around the body to the vital organs. Sufficient CO will lead to death.

3. a. $2CO_{(g)} + O_{2(g)} \longrightarrow 2CO_{2(g)}$
       $2x$ x $2x$
   b. Volume of $CO_2 + CO + N_2 = 120\ cm^3$

   Let x = volume of oxygen reacting with the CO.

Therefore, volume of O_2 in excess $= 40 - x.$

Volume of $CO_2 = 140 - 70 = 70 \text{ cm}^3$

Initial volume of $CO_2 = 70 - 2x,$

Therefore, $120 = (70 - 2x) + 2x + \text{Volume of } N_2$

Hence, volume of $N_2 = 50 \text{ cm}^3.$

Unreacted O_2 volume $= 70 - 50 = 20 \text{ cm}^3$

i.e. $x = 20 \text{cm}^3.$

Therefore, volume of CO in exhaust gas $= 40 \text{ cm}^3.$

Hence, the composition of the exhaust gases:

$$40 \text{cm}^3 \text{ CO} : 50 \text{ cm}^3 \text{ N}_2 : 30 \text{ cm}^3 \text{ CO}_2.$$

c % by volume of $CO_2 = 30 \times 100/120 = 25\%$

d $KOH_{(aq)} + CO_{2(g)} \longrightarrow KHCO_{3(aq)}$

33　Health Drink

1 Glucose:　$5 \times 60/100 = 3.0 \text{ g}$
Ascorbic acid: $1 \times 5/100 = 0.05 \text{ g}$
Phenylephrine hydrochloride: $5 \times 0.5/100 = 0.025 \text{ g}$
Paracetamol: $5 \times 10/100 = 0.5 \text{ g}$
Sodium citrate: $5 \times 15/100 = 0.75 \text{ g}$

2 Number of moles of glucose $= 0.0167$
Number of moles of paracetamol $= 0.0033$

3 a

$$\text{HOOC-CH}_2-\underset{\underset{\text{COOH}}{|}}{\overset{\overset{\text{OH}}{|}}{\text{C}}}-\text{CH}_2\text{-COOH} \quad + 3\text{NaOH}$$

$$\downarrow$$

$$\text{Na}^+ \ {}^-\text{OOC-CH}_2-\underset{\underset{\text{COO}^- \text{ Na}^+}{|}}{\overset{\overset{\text{OH}}{|}}{\text{C}}}-\text{CH}_2\text{COO}^-\text{Na}^+ \quad + 3\text{H}_2\text{O}$$

b Phenolphthalein.
c Number of moles of NaOH $= 30 \times 1/1000$
$$= 3 \times 10^{-2}$$
Therefore number of moles of citric acid $= 3 \times 10^{-2}/3$
$$= 1 \times 10^{-2}$$
Mass of citic acid $= 1 \times 10^{-2} \times 192$
$$= 1.92 \text{ g}$$

d

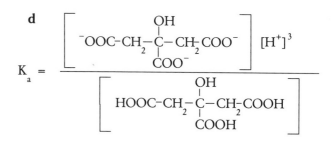

$$K_a = \frac{\left[\ {}^-\text{OOC-CH}_2-\underset{\underset{\text{COO}^-}{|}}{\overset{\overset{\text{OH}}{|}}{\text{C}}}-\text{CH}_2\text{COO}^-\ \right][\text{H}^+]^3}{\left[\ \text{HOOC-CH}_2-\underset{\underset{\text{COOH}}{|}}{\overset{\overset{\text{OH}}{|}}{\text{C}}}-\text{CH}_2\text{-COOH}\ \right]}$$

4 Glucose – for energy.
Ascorbic acid – important in the hydroxylation of collagen. Defective collagen produces the skin lesions and blood vessel weaknesses which are characteristic of scurvy.

34　Chemical Rockets

1 In rocket fuels the oxidant is the compound that provides the oxygen for combustion to take place. In this case the oxidant is N_2O_4.

2 a $250 \text{ dm}^3.$

b $N_{2(g)} = 450 \text{ dm}^3/\text{second}$
$CO_{2(g)} = 200 \text{ dm}^3/\text{second}$
$H_2O_{(g)} = 600 \text{ dm}^3/\text{second}$

3 To ensure all the methylhydrazine has been used up.

4 If the temperature was increased, the equilibrium would move to the right as it will favour the endothermic process.

5 If the pressure was increased the equilibrium would move to the left as it will favour the side with the fewest moles of gas. This has the effect of decreasing the pressure.

6 It has reactions with other compounds in the environment which lead to adverse effects on plants, animals and the general environment.

35 North Sea Analysis

1 a $MgSO_4$ **b** $CaCO_3$ **c** KCl **d** KBr **e** $CaSO_4$.

2 a Number of moles of $Cl^-_{(aq)}$ present:
(i) in $MgCl_2$
Number of moles of $MgCl_2 = 2.73/95$
$$= 0.029$$
Therefore, number of moles of $Cl^- = 2 \times 0.029$
$$= 0.058$$

(ii) in KCl

Number of moles of KCl = 0.70/74.5

                      = 0.0094

Therefore, number of moles of Cl^- = 0.0094

(iii) in NaCl

Number of moles of NaCl = 27.40/58.5

                      = 0.468

Therefore, number of moles of Cl^- = 0.468

Therefore, the total number of moles of Cl^-

                      = 0.5354 moles in 1 dm^3.

Hence, in 25 cm^3 of solution there will be

0.5354/40 moles of Cl^-

= 0.0133 moles of Cl^- ions.

$$Ag^+_{(aq)} + Cl^-_{(aq)} \longrightarrow AgCl_{(s)}$$

Hence, number of moles of $Ag^+_{(aq)}$ = 0.0133

Volume of $AgNO_3$ = 13.3 cm^3.

b Potassium chromate(VI), K_2CrO_4.

White precipitate (AgCl) to pink precipitate (Ag_2CrO_4).

3 a i Calcium carbonate, $CaCO_3$

  ii Carbon dioxide, CO_2.

b $$CaCO_{3(s)} \longrightarrow CaO_{(s)} + CO_{2(g)}$$

c Number of grams of $CaCO_3$ = 0.14 g

Number of moles of $CaCO_3$ = 0.14/100

                      = 1.4×10^{-3}

Therefore, number of moles of $CO_{2(g)}$ lost

                      = 1.4×10^{-3}

Mass of $CO_{2(g)}$ lost = $1.4 \times 10^{-3} \times 44$

                    = 0.0616g

4 a i $$Mg^{2+}_{(aq)} + 2OH^-_{(aq)} \longrightarrow Mg(OH)_{2(s)}$$

  ii $$Mg(OH)_{2(s)} + 2HCl_{(aq)} \longrightarrow MgCl_{2(aq)} + 2H_2O_{(l)}$$

b Anode: $2Cl^- \longrightarrow Cl_2 + 2e^-$

Cathode: $Mg^{2+} + 2e^- \longrightarrow Mg$

---

36   Water

1 K_W increases with increase in temperature.

2 Positive enthalpy change indicated by increased value of K_W reflecting increased concentration of hydrogen and hydroxide ions due to equilibrium moving to the right.

3 For pure water $[H^+_{(aq)}] = [OH^-_{(aq)}]$

               $[H^+_{(aq)}] = 2.34 \times 10^{-7}$ mol dm^{-3}

                   pH = $-\log_{10}(2.34 \times 10^{-7})$

                         = 6.63.

4 Carbon dioxide dissolving in water forming carbonic acid, H_2CO_3.

$$CO_{2(g)} + H_2O_{(l)} \rightleftharpoons H_2CO_{3(aq)}$$

Carbonic acid is a weak acid and hence will reduce the pH value.

5 a Acid rain, formed from the reaction between $SO_{2(g)}$ from the combustion of fossil fuels containing sulphur, and water forming sulphurous acid and subsequent oxidation of sulphuric acid.

b i pH = $-\log_{10}[H+_{(aq)}]$

$[H+_{(aq)}] = 1 \times 10^{-4}$ mol dm^{-3}

Therefore, volume of NaOH required

                = $(1 \times 10^{-4}/10) \times 1000/0.001$

                = 10 cm^3.

  ii Phenolphthalein. Colourless to pink solution.

c $$H^+_{(aq)} + CO_3^{2-}{}_{(s)} \rightleftharpoons HCO_3^-{}_{(aq)}$$

---

37   Chlorine in the Environment

1 +1

2 a $$Cl_{2(g)} + H_2O_{(l)} \longrightarrow HClO_{(aq)} + HCl_{(aq)}$$

b Chlorine is a very poisonous gas and would undoubtedly kill marine life generally and be dangerous to the nearby environment.

3 a $$2ClO^-_{(aq)} \xrightarrow{\text{sunlight}} 2Cl^-_{(aq)} + O_{2(g)}$$

b Measure the rate at which the oxygen gas was produced by collecting the gas in a gas syringe.

4 a $$2NaOH_{(aq)} + Cl_{2(g)} \longrightarrow NaClO_{(aq)} + NaCl_{(aq)} + H_2O_{(l)}$$

b Simultaneous oxidation and reduction.

In going from Cl_2 to ClO^- the chlorine has been oxidised, from Cl_2 to Cl^- it has been reduced.

5 a NaClO= +1   NaCl= −1 $NaClO_3$ = +5

b Disproportionation.

c NaClO−bleaches.

NaCl−flavouring food.

$NaClO_3$ −weedkiller.

38 Forensic Science

1 Number of moles of CO_2 = 0.0435
Number of moles of carbon = 0.0435
Therefore, mass of carbon = 0.522 g

Number of moles of H_2O = 0.0652
Number of moles of hydrogen = 0.1304
Therefore, mass of hydrogen = 0.1304 g
Mass of oxygen in 1 g of this compound
= 1 − (0.522 + 0.1304)
= 0.3476 g
Empirical formula is C_2H_6O

2 Last 'stick' on the mass spectrum shows the molecular mass to be 46 (due to the detection of $C_2H_6O^+$ ion).

3 **a** Molecular formula: C_2H_6O.

b CH_3-O-CH_3 and CH_3CH_2OH.

4 Structure of the substance is CH_3CH_2OH.

Reasons – Strong absorption around 3350 cm^{-1} indicating O−H...O (H−bond) stretch. CH_3CH_2OH would have hydrogen bonding between molecules. In addition, there is strong absorption around 1025−1110 cm^{-1} indicating a C−OH stretch.

5 Iodoform reaction (triiodomethane reaction).
Reagents – I_2/OH^- which is a mild oxidant. The mild oxidation changes ethanol to ethanal. The iodoform reaction indicates the presence of the $CH_3C=O$ group which the oxidised ethanol now contains. A positive test is indicated by the formation of a yellow precipitate of CHI_3.

39 Lead Pollution

1 Number of moles of TEL added to 1 dm^3 of petrol
= 1.3/323 = 0.004

Therefore, mass of lead = 207 x 0.004 = 0.833 g.

2 Number of dm^3 of petrol used = 15000/9 = 1666.7 .

Therefore, mass of lead = 1666.7 x 0.833
= 1388.36 g (1.388 kg).

3 Number of moles of lead in 0.15 g = 0.15/207
= 7.25 x 10^{-4}

Therefore, mass of TEL now allowed = 7.25 x 10^{-4} x 323
= 0.234 g dm^{-3}.

4 **a** Number of moles of lead = 10000 x 106/207
= 4.83 x 10^7

Therefore, mass of TEL added to petrol per year
= 4.83 x 10^7 x 323 = 1.56 x 1010 g

= 1.56 x 10^7 kg.

b Number of moles of chloroethane required
= 4 x 4.83 x 10^7 = 1.932 x 10^8

Mass of chloroethane = 1.932 x 10^8 x 64.5
= 1.25 x 10^7 kg

5 **a** Electrophilic addition.

b
$$H_2C{=}CH_2 \longrightarrow H_3C{-}\overset{+}{C}H_2 \longrightarrow H_3C{-}CH_2Cl$$
$H\,\delta{+}$
Cl^-
$Cl\,\delta{-}$

40 Corrosion

1 **a** A REDOX process relates to the processes of oxidation and reduction, which are intimately connected in that during oxidation the oxidising agent itself becomes reduced and vice versa. Therefore, an oxidation process is always accompanied by a reduction process.

b $Fe_{(s)} \longrightarrow Fe^{2+}_{(aq)} + 2e^-$

c $2Fe_{(s)} + O_{2(g)} + 2H_2O_{(l)} \longrightarrow 2Fe^{2+}_{(aq)} + 4OH^-_{(aq)}$

2 **a** Fe^{2+} [Ar] $3d^6$: Fe^{3+} [Ar] $3d^5$

b Oxidation occurs due to the oxygen in the air. Fe^{3+} has a more stable electronic configuration.

3 **a** Mg and Zn have a more negative standard electrode potential than iron, i.e. would protect.

b Sn and Cu have a less negative standard electrode potential than iron, i.e. would not protect.

4 To act as a source of zinc ions which can make use of the advantageous Zn^{2+}/Zn standard electrode potential.

5 **a** Nitrate ion.

b Dative covalent and covalent.

c $[Fe(NO_3)_4]^{2+}$

d Zinc chromate(VI) primer.

21 Enthalpy of Combustion

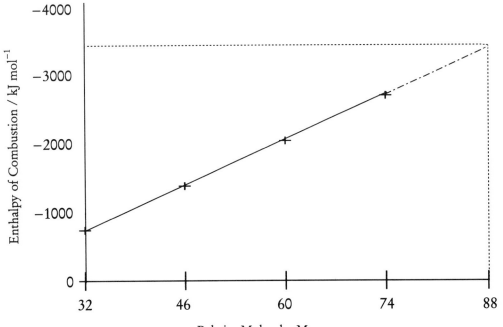

22 Reaction Kinetics

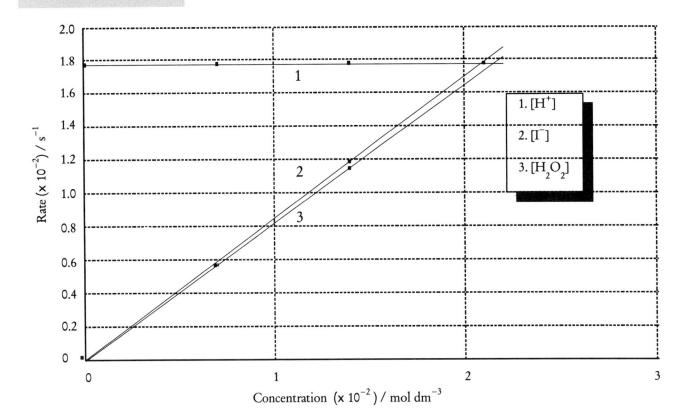

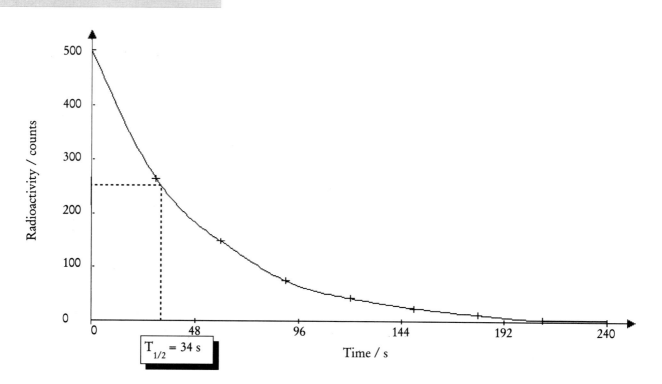

PERIODIC TABLE

Group	1 (I)	2 (II)		3 (III)	4 (IV)	5 (V)	6 (VI)	7 (VII)	0
1									4 He 2 Helium
	7 Li 3 Lithium	9 Be 4 Beryllium		11 B 5 Boron	12 C 6 Carbon	14 N 7 Nitrogen	16 O 8 Oxygen	19 F 9 Fluorine	20 Ne 10 Neon
2	23 Na 11 Sodium	24 Mg 12 Magnesium		27 Al 13 Aluminium	28 Si 14 Silicon	31 P 15 Phosphorus	32 S 16 Sulphur	35.5 Cl 17 Chlorine	40 Ar 18 Argon
3	39 K 19 Potassium	40 Ca 20 Calcium							

Hydrogen — 1 H 1

Transition metals

39 K 19 Potassium	40 Ca 20 Calcium	45 Sc 21 Scandium	48 Ti 22 Titanium	51 V 23 Vanadium	52 Cr 24 Chromium	55 Mn 25 Manganese	56 Fe 26 Iron	59 Co 27 Cobalt	59 Ni 28 Nickel	63.5 Cu 29 Copper	65 Zn 30 Zinc	70 Ga 31 Gallium	73 Ge 32 Germanium	75 As 33 Arsenic	79 Se 34 Selenium	80 Br 35 Bromine	84 Kr 36 Krypton
85 Rb 37 Rubidium	88 Sr 38 Strontium	89 Y 39 Yttrium	91 Zr 40 Zirconium	93 Nb 41 Niobium	96 Mo 42 Molybdenum	99 Tc 43 Technetium	101 Ru 44 Ruthenium	103 Rh 45 Rhodium	106 Pd 46 Palladium	108 Ag 47 Silver	112 Cd 48 Cadmium	115 In 49 Indium	119 Sn 50 Tin	122 Sb 51 Antimony	128 Te 52 Tellurium	127 I 53 Iodine	131 Xe 54 Xenon
133 Cs 55 Caesium	137 Ba 56 Barium	178.5 Hf 72 Hafnium	181 Ta 73 Tantalum	184 W 74 Tungsten	186 Re 75 Rhenium	190 Os 76 Osmium	192 Ir 77 Iridium	195 Pt 78 Platinum	197 Au 79 Gold	201 Hg 80 Mercury	204 Tl 81 Thallium	207 Pb 82 Lead	209 Bi 83 Bismuth	209 Po 84 Polonium	210 At 85 Astatine	222 Rn 86 Radon	
223 Fr 87 Francium	226 Ra 88 Radium	261 Unq 104 Unnil-quadium	262 Unp 105 Unnil-pentium	263 Unh 106 Unnil-hexium	262 Uns 107 Unnil-septium	Uno 108 Unnil-octium	Une 109 Unnil-ennium										

Lanthanide series

139 La 57 Lanthanum	140 Ce 58 Cerium	141 Pr 59 Praseodymium	144 Nd 60 Neodymium	147 Pm 61 Promethium	150 Sm 62 Samarium	152 Eu 63 Europium	157 Gd 64 Gadolinium	159 Tb 65 Terbium	162 Dy 66 Dysprosium	165 Ho 67 Holmium	167 Er 68 Erbium	169 Tm 69 Thulium	173 Yb 70 Ytterbium	175 Lu 71 Lutetium
227 Ac 89 Actinium	232 Th 90 Thorium	231 Pa 91 Protactinium	238 U 92 Uranium	237 Np 93 Neptunium	244 Pu 94 Plutonium	243 Am 95 Americium	247 Cm 96 Curium	247 Bk 97 Berkelium	251 Cf 98 Californium	252 Es 99 Einsteinium	257 Fm 100 Fermium	258 Md 101 Mendelevium	259 No 102 Nobelium	260 Lw 103 Lawrencium